AF744140

Nuclear Fusion - One Noble Goal and a Variety of Scientific and Technological Challenges

Edited by Igor Girka

Published in London, United Kingdom

IntechOpen

Supporting open minds since 2005

Nuclear Fusion - One Noble Goal and a Variety of Scientific and Technological Challenges
http://dx.doi.org/10.5772/intechopen.75348
Edited by Igor Girka

Contributors
Fouad A. Majeed, Yousif A Abdul-Hussien, Fatima M. Hussian, Rong Liu, Zhang Xingyi, Elena Koresheva, Irina Aleksandrova, Andrei Nikitenko, Eugeniy Koshelev, Boris Kuteev, Quamrul Haider

© The Editor(s) and the Author(s) 2019
The rights of the editor(s) and the author(s) have been asserted in accordance with the Copyright, Designs and Patents Act 1988. All rights to the book as a whole are reserved by INTECHOPEN LIMITED. The book as a whole (compilation) cannot be reproduced, distributed or used for commercial or non-commercial purposes without INTECHOPEN LIMITED's written permission. Enquiries concerning the use of the book should be directed to INTECHOPEN LIMITED rights and permissions department (permissions@intechopen.com).
Violations are liable to prosecution under the governing Copyright Law.

Individual chapters of this publication are distributed under the terms of the Creative Commons Attribution 3.0 Unported License which permits commercial use, distribution and reproduction of the individual chapters, provided the original author(s) and source publication are appropriately acknowledged. If so indicated, certain images may not be included under the Creative Commons license. In such cases users will need to obtain permission from the license holder to reproduce the material. More details and guidelines concerning content reuse and adaptation can be found at http://www.intechopen.com/copyright-policy.html.

Notice
Statements and opinions expressed in the chapters are these of the individual contributors and not necessarily those of the editors or publisher. No responsibility is accepted for the accuracy of information contained in the published chapters. The publisher assumes no responsibility for any damage or injury to persons or property arising out of the use of any materials, instructions, methods or ideas contained in the book.

First published in London, United Kingdom, 2019 by IntechOpen
IntechOpen is the global imprint of INTECHOPEN LIMITED, registered in England and Wales, registration number: 11086078, The Shard, 25th floor, 32 London Bridge Street
London, SE19SG - United Kingdom
Printed in Croatia

British Library Cataloguing-in-Publication Data
A catalogue record for this book is available from the British Library

Additional hard copies can be obtained from orders@intechopen.com

Nuclear Fusion - One Noble Goal and a Variety of Scientific and Technological Challenges
Edited by Igor Girka
p. cm.
Print ISBN 978-1-78985-787-0
Online ISBN 978-1-78985-788-7

For EU product safety concerns:
IN TECH d.o.o., Prolaz Marije Krucifikse Kozulić 3, 51000 Rijeka, Croatia, info@intechopen.com or visit our website at intechopen.com.

We are IntechOpen,
the world's leading publisher of
Open Access books

Built by scientists, for scientists

4,100+
Open access books available

116,000+
International authors and editors

120M+
Downloads

151
Countries delivered to

Our authors are among the
Top 1%
most cited scientists

12.2%
Contributors from top 500 universities

CLARIVATE ANALYTICS BOOK CITATION INDEX INDEXED

WEB OF SCIENCE™

Selection of our books indexed in the Book Citation Index
in Web of Science™ Core Collection (BKCI)

Interested in publishing with us?
Contact book.department@intechopen.com

Numbers displayed above are based on latest data collected.
For more information visit www.intechopen.com

Meet the editor

Igor O. Girka earned his MS degree in Theoretical Nuclear Physics and Candidate of Science and Doctor of Science degrees in Plasma Physics from V.N.Karazin Kharkiv National University (KKhNU), Kharkiv, Ukraine, in 1985, 1991, and 2004, respectively. In 2003–2005, he was a vice president of KKhNU. Since 2005, he has been the dean of the KKhNU-School of Physics and Technology. He has authored 5 textbooks, 3 monographs, and 99 papers. His current research interests include propagation, conversion, and absorption of magnetohydrodynamic and surface-type electromagnetic waves in fusion devices and devices for plasma electronics. Since 2015, he has been the vice president of the Ukrainian Physical Society. In 2018, he was elected a corresponding member of the National Academy of Sciences of Ukraine.

Contents

Preface

Power production and its consumption and distribution are among the most urgent problems of civilization. Despite huge efforts and positive dynamics in introducing renewable sources of energy such as solar and wind, nuclear power plants still remain the major source of carbon-free electric energy. However, the deposits of nuclear fuel are limited. Fusion can be an alternative to fission for the foreseeable future. Nuclear fusion takes place in the sun and provides energy for life on Earth. Research in the field of controlled nuclear fusion has been ongoing for almost 100 years. Since 1920, many physics and engineering problems of fusion have been successfully resolved, and several fusion technologies have been implemented in other fields of science and technology. Magnetic confinement systems are the most promising for effective implementation, and the International Thermonuclear Experimental Reactor (ITER) is under construction in France. ITER is designed to demonstrate a 10-fold return on energy.

To accomplish nuclear fusion on Earth, we have to resolve a wide scope of scientific and technological problems. This is why the nuclear fusion international community consists of a large number of divisions. For example, nuclear fusion problems start from those of nuclear physics, which demonstrate the basics of energy release from the fusion of light atoms to heavier atoms. To realize controlled nuclear fusion, many schemes have been discovered and suggested, for example magnetic confinement and inertial synthesis. Magnetic confinement systems can be divided into tokamaks and stellarators. These two types of nuclear fusion devices can in turn be classified into spheromaks and torsatrons. Plasma is proposed for creating the conditions for controlled nuclear fusion. Plasma physics and engineering are also a very large part of physics. Plasma used to be unstable and needed additional efforts to prevent plasma discharge breakdown. Many physicists have devoted themselves to studying plasma instabilities and searching for ways to suppress or avoid these instabilities. Electromagnetic waves propagate in the fusion plasmas. The waves can be used for plasma heating up to the temperatures of nuclear fusion reactions. We also have to understand how to excite these waves and how to propagate them. These waves interact with the plasma, the walls of the chamber, and with each other, and we need to understand how they are absorbed. Many problems are associated with designing and constructing the antennae to excite electromagnetic waves and the power supply for the antennae. We have to know how to extract the energy from the future thermonuclear power plant. Finally, we need materials to produce a chamber in which the plasmas can be "boiled." These materials should be stable to significant heat and mechanical loads, as well as to undesirable interaction with aggressive hot plasmas.

For decades, many national and international journals have published papers devoted to fusion-related topics. It is clear that any single book cannot cover all the topics of nuclear fusion research. This monograph includes selected chapters of nuclear physics and mechanical engineering within the scope of nuclear fusion.

In the first chapter, "Nuclear Fusion: Holy Grail of Energy," harnessing the energy produced in a nuclear fusion reaction in a laboratory environment is discussed.

Various research programs in the field of controlled nuclear fusion are also discussed. Emphasis is given to overcoming some of the technological challenges, such as surmounting the Coulomb barrier, confining the plasma, and achieving the ignition temperature.

The second chapter, "Fusion Reaction of Weakly Bound Nuclei," is devoted to comparison of a semiclassical and full quantum mechanical approach to study the total fusion reaction cross-section and the fusion barrier distribution for several systems such as $^{6}Li + ^{64}Ni$, $^{11}B + ^{159}Tb$, and $^{12}C + ^{9}Be$. The results obtained from the numerical calculations based on these two approaches are compared with the available experimental data.

Thorium is a fertile element that can be applied in the conceptual blanket design of a fusion/fission hybrid energy reactor, in which ^{232}Th is mainly used to breed ^{233}U by capture reaction. The activation γ-ray offline method for determining the thorium reaction rates is developed in the third chapter, "Fusion Neutronics Experiments for Thorium Assemblies" The ^{232}Th (n, γ), ^{232}Th (n, f), and ^{232}Th (n, 2n) reaction rates in assemblies are measured by using ThO_2 foils and an HPGe γ spectrometer.

The fourth chapter, "Mechanical Mockup of IFE Reactor Intended for the Development of Cryogenic Target Mass Production and Target Rep-Rate Delivery into the Reaction Chamber" describes the efforts that are underway at the Lebedev Physical Institute of Russian Academy of Sciences to arrange target production and delivery into the reaction chamber for inertial fusion. The current status and future trends of developments in the area of advanced target technologies are discussed.

While fusion plasmas are very hot, magnetic coils should be very cold to provide stable static magnetic fields of high intensity. The mechanical behavior of the cable in a conduit conductor is studied in the fifth chapter, "The Mechanical Behaviors of Cable-in-Conduit Conductor for the ITER Project," because of its high importance for understanding the mechanical response and assessing the safety of the superconducting structures in the ITER.

This book is intended for nuclear scientists, engineers, and advanced students with a basic understanding of nuclear fusion and power reactors. It discusses several issues of nuclear physics as well as those of mechanical engineering related to nuclear fusion.

Igor Girka, Professor
Corresponding Member of National Academy of Sciences of Ukraine,
V. N. Karazin Kharkiv National University, Ukraine

Section 1

Selected Problems of Nuclear Physics

Chapter 1

Nuclear Fusion: Holy Grail of Energy

Quamrul Haider

Abstract

The declining reserves of fossil fuels and their detrimental effects on the environment have thrust nuclear power based on fission reaction into the limelight as a promising option to energy-starved economies around the world. However, the 1986 Chernobyl and 2011 Fukushima accidents have heightened our fears about nuclear technology's ability to provide a safe way of generating clean power. There is another kind of nuclear energy that has been powering the Sun and stars since their formation. It is nuclear fusion—a process in which two lighter nuclei, typically isotopes of hydrogen, combine together under conditions of extreme pressure and temperature to form a heavier nucleus. In this chapter, harnessing the energy produced in nuclear fusion reaction in a laboratory environment is discussed. Various research programs dedicated to building fusion reactors are also discussed. Emphasis is given on overcoming some of the technological challenges, such as surmounting the Coulomb barrier, confining the plasma, and achieving the "ignition" temperature for fusion.

Keywords: nuclear fission and fusion, fission and fusion reactors, fusion in the Sun, fusion on Earth, cold fusion, Coulomb barrier, fusion "ignition" temperature, Lawson criterion, Debye length, plasma confinement, magnetic and inertial confinements, tokamak, stellarators, fusion torch

1. Introduction

The 1930s were heady times for nuclear physics. A "hit parade" of discoveries gave us new insights into the properties of the nucleus. The means of unlocking enormous amount of energy stored inside a nucleus seemed at hand. Finally, the discovery of nuclear fission in 1938 ushered in a new era in the history of mankind —the nuclear age [1].

Nuclear energy is a technologically proven nonfossil energy source that made significant contribution to the world's energy supply in the past 6 decades. There are two nuclear processes in which enormous amount of energy is released from nuclear bonds between the particles within the nucleus. They are nuclear fission and nuclear fusion.

2. Nuclear fission

The importance of nuclear fission for the production of energy is obvious. In fission reactions, a heavy nucleus is split into two lighter fragments and two or three neutrons. About 180 MeV of energy is produced in the fission of an actinide to one of its most probable daughter pairs. This means that 1 kg of uranium (^{235}U) is capable of producing enough energy to keep a 100-Watt light bulb running for about 25,000 years [2].

2.1 Fission reactors

All nuclear power plants in operation today rely on controlled fission of the isotopes of uranium and plutonium [3]. The reactor functions primarily as an exotic heat source to turn water into pressurized steam. Only the source of heat energy differs—nuclear power plants use fissile radioactive nucleus, while nonnuclear power plants use fossil fuel. The rest of the power train is the same. The steam turns the turbine blades, the blades generate mechanical energy, the energy runs the generator, and the generator produces electricity. The major improvement is the elimination of the combustion products of fossil fuels—the greenhouse gases, which have destroyed our environment beyond repair.

Because of its abundance in nature, most nuclear reactors use uranium as fuel. Natural uranium contains 0.7% of the fissile ^{235}U; the rest is non-fissile ^{238}U. When ^{235}U is bombarded with a slow neutron, it captures the neutron to form ^{236}U, which undergoes fission producing two lighter fragments and releases energy together with two or three neutrons. The neutrons produced in the reaction cause more fission resulting in a self-sustaining chain reaction. A reactor is considered safe when a self-sustained chain reaction is maintained with exactly one neutron from each fission inducing yet another fission reaction.

2.2 Problems and concerns with fission reactors

Although fission-based nuclear reactors generate huge amounts of electricity with zero greenhouse gas emissions, and thus was hailed as a solution not only to global warming but also to global energy needs, nuclear energy is now seen by many, and with good reasons, as the misbegotten stepchild of nuclear weapons programs. Besides, it is by no means certain that the safety systems designed to shut down the reactor in the event of a runaway reaction are 100% foolproof and will work as designed. Another area of great concern is the hazards associated with the disposal of highly radioactive waste products.

What has raised our fear in regard to nuclear power more than anything else are the accidents at Chernobyl in 1986 and Fukushima in 2011. They were a sobering reminder of what we can expect from an accident due to catastrophic reactor failure or human errors. The Fukushima disaster in particular has shattered the zero risk myth of power reactors and heightened our concern about the invisibility of the added lethal component, nuclear radiation. Consequently, they have spurred our interest in the other source of nuclear energy—fusion.

3. Nuclear fusion

Nuclear fusion is the process in which two lighter nuclei, typically isotopes of hydrogen, combine together under conditions of extreme pressure and temperature to form a heavier nucleus, resulting in the release of enormous amount of energy. The fusion of four protons to form the helium nucleus ^{4}He, two positrons, and two neutrinos, for example, generates about 27 MeV of energy.

In the 1930s, scientists, particularly Hans Bethe, discovered that it is fusion that has been powering the Sun and stars since their formation [4]. A "fusion reactor" buried deep in the Sun's interior produces in one heartbeat the energy of 100 billion nuclear bombs. Beginning in the 1940s, researchers began to look for ways to initiate and control fusion reactions to produce useful energy on Earth. We now have a very good understanding of how and under what conditions two nuclei can fuse together.

3.1 Fusion in the Sun

The fusion of hydrogen into helium in the Sun and other stars occurs in three stages. First, two ordinary hydrogen nuclei (^{1}H), which are actually just single protons, fuse to form an isotope of hydrogen called deuterium (^{2}H), which contains one proton and one neutron. A positron (e^+) and a neutrino (ν) are also produced. The positron is very quickly annihilated in the collision with an electron, and the neutrino travels right out of the Sun:

$$^1H + {}^1H \rightarrow {}^2H + e^+ + \nu. \tag{1}$$

Once created, the deuterium fuses with yet another hydrogen nucleus to produce ^{3}He—an isotope of ^{4}He. At the same time, a high-energy photon, or γ ray, is produced. The reaction is

$$^2H + {}^1H \rightarrow {}^3He + \gamma. \tag{2}$$

The final step in the reaction chain, which is called the proton-proton cycle, takes place when a second ^{3}He nucleus, created in the same way as the first, collides and fuses with another ^{3}He, forming ^{4}He and two protons. In symbols,

$$^3He + {}^3He \rightarrow {}^4He + 2^1H. \tag{3}$$

The net result of the proton-proton cycle is that four hydrogen nuclei combine to create one helium nucleus. The mass of the end product is 0.0475×10^{-27} kg less than the combined mass of the ^{3}He nuclei. This mass difference, known as mass defect in the parlance of nuclear physics, is converted into 26.7 MeV of energy as known from Einstein's equation $E = mc^2$.

The proton-proton cycle is particularly slow—only one collision in about 10^{26} for the cycle to start. As the cycle proceeds, the Sun's temperature rises, and eventually three ^{4}He nuclei combine to produce ^{12}C. Despite the slowness of the proton-proton cycle, it is the main source of energy for the Sun and for stars less massive than the Sun. The amount of energy released is enough to keep the Sun shining for billions of years.

Besides the proton-proton cycle, there is another important set of hydrogen-burning reactions called the carbon-nitrogen-oxygen (CNO) cycle that occurs at higher temperatures. Although CNO cycle contributes only a small amount to the Sun's luminosity, it dominates in stars that are more massive than a few times the Sun's mass. A star like Sirius with somewhat more than twice the mass of the Sun derives almost all of its energy from the CNO cycle.

4. Coulomb barrier

An obstacle called the Coulomb barrier caused by the strongly repulsive electrostatic forces between the positively charged nuclei prevents them from fusing under normal circumstances. However, fusion can occur under conditions of extreme pressure and temperature. That is why fusion reaction is often termed as thermonuclear reaction.

Nuclei, which have positive charges, must collide at extremely high speeds to overcome the Coulomb barrier. The speed of particles in a gas is governed by the temperature. At the very center of the Sun and other stars, it is extremely hot and density is very high. For the Sun, the temperature is around 15 million degrees Celsius, and the central density is about 150 times that of water. Under such extreme conditions, electrons in an atom become completely detached from the atomic nucleus, thereby forming an ionized fluid called plasma—a "soup" of hot gas, with bare, positively charged atomic nuclei and negatively charged electrons whizzing about at extremely high speeds. The plasma as a mixture of positive ions (nuclei) and negative electrons is overall electrically neutral.

Without the high pressure of the overlying layers, the hot plasma at the solar core would simply explode into space, shutting off the nuclear reactions. The pressure, which is about 250 billion atmospheres at the Sun's core, squeezes the nuclei so that they are within 1 fm (10^{-15} m) of each other. At this distance, the attractive strong nuclear force that binds protons and neutrons together in the nucleus becomes dominant and pulls the incoming particles together, causing them to fuse.

Additionally, massive gravitational force causes nuclei to be crowded together very densely. This means collisions occur very frequently, another requirement if a high fusion rate is to occur. A quick and crude calculation suggests that we need about 10^{38} collisions per second to keep the Sun going, while within the core we get about 10^{64} collisions per interactions per second, implying only one in 10^{26} collisions needs to be a successful fusion event.

5. Nuclear fusion on Earth

One of the major challenges in initiating a fusion reaction in a laboratory environment on Earth is to create conditions similar to that in the Sun—extremely high temperatures, perhaps more than 100 million degrees Celsius (equivalent to mean particle kinetic energies of ~10 keV) while simultaneously maintaining a high enough density for a long enough time so that the rate of fusion reactions will be large enough to generate the desired power.

5.1 "Ignition" temperature

We can estimate the minimum temperature required to initiate fusion by calculating the Coulomb barrier which opposes two protons approaching each other to fuse. With $e^2 = 1.44$ MeV-fm, where e is the charge of a proton, and $r = 1.0$ fm (separation between two protons), the height of the Coulomb barrier is

$$U = \frac{e^2}{r} = 1.44 \ \text{MeV}. \tag{4}$$

The kinetic energy of the nuclei moving with a speed v is related to the temperature T by

$$\frac{1}{2}mv^2 = \frac{3}{2}\ k_B T, \tag{5}$$

where $k_B = 8.62 \times 10^{-11}$ MeV/K is the Boltzmann constant. By equating the average thermal energy to the Coulomb barrier height and solving for T gives a value for the temperature of around 10 billion Kelvin (K).

The above "back of the envelop" calculation, using classical physics, does not take into consideration the quantum effect of tunneling, which predicts there will be a small probability that the Coulomb barrier will be overcome by nuclei tunneling through it. The probability P of such an event happening is

$$P \propto \exp\left(-\frac{Z_1 Z_2 \alpha c}{v}\right), \tag{6}$$

where Z_1 and Z_2 are the atomic numbers of the interacting particles, $\alpha = 1/137$ is the fine structure constant, and v is the relative velocity of the colliding nuclei. Using the classical turning point r_0 and de Broglie wavelength $\lambda = h/p = h/mv$, the above expression can be written as

$$P \propto \exp\left(-\frac{r_0}{\lambda}\right). \tag{7}$$

Thus, large values of v (high energies), or small λ, favor a fusion reaction.

Taking into account the tunneling probability, we can now estimate the temperature for fusion to occur. In terms of de Broglie wavelength, the kinetic energy is

$$K = \frac{p^2}{2m} = \frac{h^2}{2m\lambda^2}. \tag{8}$$

If we require that the nuclei must be closer than the de Broglie wavelength for tunneling to take over and the nuclei to fuse, then the Coulomb barrier is given by

$$\frac{e^2}{\lambda} = \frac{h^2}{2m\lambda^2} \rightarrow \lambda = \frac{h^2}{2me^2}. \tag{9}$$

If we use this wavelength as the distance of closest approach to calculate the temperature, we obtain

$$\frac{3}{2}\, k_B T = \frac{e^2}{\lambda} = \frac{2me^4}{h^2}. \tag{10}$$

Solving for the temperature, we get

$$T = \frac{4me^4}{3h^2 k_B} = \frac{mc^2\alpha^2}{3\pi^2 k_B}. \tag{11}$$

For two hydrogen nuclei ($mc^2 = 940$ MeV), this gives a temperature of about 20 million Kelvin.

6. Fusion reactor

Since the 1950s, scientists have been working tirelessly to develop a reactor in order to harness the nearly inexhaustible energy produced during fusion [5]. The goals of fusion research at present include the following:

1. To achieve the required temperature to ignite the fusion reaction.

2. To keep the plasma together at this temperature long enough to get useful amounts of energy out of the thermonuclear fusion reactions.

3. To obtain more energy from the thermonuclear reactions than is used to heat the plasma to the ignition temperature.

To date, much headway has been made toward achieving these goals.

6.1 Fuel

Just like the Sun, the fuel for a fusion reactor is hydrogen, the most abundant element in the Universe. But without the benefit of gravitational force that is at work in the Sun, achieving fusion on Earth requires a different approach. The simplest reaction in which enormous amount of energy will be released is the fusion of the hydrogen isotopes deuterium (^2H) and tritium (^3H) producing ^4He and a neutron. For the sake of brevity, we will use the notation d and t for deuterium and tritium, respectively.

Deuterium is found aplenty in ocean water, enough to last for billions of years. This makes it an attractive source of alternative energy relative to other sources of energy. Naturally occurring tritium, on the other hand, is extremely rare. It is radioactive with a half-life of around 12 years. Trace quantities of tritium can be found in cosmic rays. Nevertheless, it can be produced inside a reactor by neutron (n) activation of lithium (Li), the other raw material for fusion found in brines, minerals, and clays. Because of the abundance of fusion fuel, the amount of energy that can be released in controlled fusion reactions is virtually unlimited.

For d-t reaction, we must first create the tritium from either flavor of lithium:

$$^6\mathrm{Li} + \mathrm{n} \rightarrow {}^4\mathrm{He} + \mathrm{t}, \tag{12}$$

or

$$^7\mathrm{Li} + \mathrm{n} \rightarrow {}^4\mathrm{He} + {}^4\mathrm{H} \rightarrow {}^4\mathrm{He} + \mathrm{t} + \mathrm{n}. \tag{13}$$

The next step in the reaction is

$$\mathrm{d} + \mathrm{t} \rightarrow {}^5\mathrm{He} \rightarrow {}^4\mathrm{He} + \mathrm{n}. \tag{14}$$

The neutrons generated from the d-t fusion can be used to bombard lithium to produce helium and tritium, thereby starting a controlled, sustainable chain reaction.

The mass of the resulting helium atom and neutron is not the exact sum of the masses of deuterium and tritium. Once again, because of mass defect, each lithium nucleus converted to tritium will end up yielding about 18 MeV of thermal energy. Compared to fission, where each split of uranium releases about 200 MeV of energy, it might appear that the energy released during fusion is rather small. The discrepancy in the energies lies in the number of nucleons involved in the reactions —more than 200 for fission and 5 for fusion. On a per nucleon basis, fusion releases 18/5 = 3.6 MeV, while fission releases 200/236 = 0.85 MeV. So, fusion wins hands down, by greater than a factor of 4.

The other fusion scheme for which the required fuel ($^4\mathrm{He}$) will be produced is $\mathrm{d} + \mathrm{d} \rightarrow {}^4\mathrm{He}$. Another reaction, $^2\mathrm{H} + {}^3\mathrm{He} \rightarrow {}^4\mathrm{He} + p$, is an example of a fusion reaction that releases its energy entirely in the form of charged particles, rather than neutrons, thereby offering the possibility, at least in principle, of direct conversion of fusion energy into electrical energy. However, the cross sections and reaction rates for both the reactions are as much as a factor of 10 lower than the d-t reaction. Moreover, because of the higher Coulomb barrier (~2.88 MeV), the ignition temperatures required for $^2H + {}^3He$ reaction are much higher than those of d-t fusion.

An interesting fusion reaction is a proton colliding with boron (B). The proton fuses with ^{11}B to form ^{12}C which immediately decays into three alpha (^4H nucleus) particles. A total energy of 8.7 MeV is released in the form of kinetic energy of the alpha particles. Since it is relatively easy to control the energy of the proton with today's accelerator technology, this fusion reaction can be easily initiated without involving other interaction channels.

6.2 Conditions for fusion reaction

In order to attain the temperature for fusion to occur, the plasma has to meet some conditions. They are Lawson criterion and Debye length.

6.2.1 Lawson criterion

In addition to providing a sufficiently high temperature to enable the particles to overcome the Coulomb barrier, a critical density of the ions in the plasma must be maintained to make the probability of fusion high enough to achieve a net yield of energy from the reaction. The condition which must be met for a yield of more energy than is required for the heating of the plasma is stated in terms of the product of the plasma density (n_d) and confinement time (τ). The product has to satisfy the inequality:

$$n_d\tau \geq 3 \times 10^{20}\ \text{s/m}^3. \quad (15)$$

This relation is called the Lawson criterion [6]. Researchers sometimes use the triple product of n_d, τ, and the plasma temperature T. Called the *fusion product*, the condition for fusion to take place is

$$n_d\tau T \geq 5 \times 10^{21}\ \text{s. keV/m}^3. \quad (16)$$

To summarize, three main conditions are necessary for nuclear fusion:

1. The temperature must be hot enough to allow the ions to overcome the Coulomb barrier and fuse together. This requires a temperature of at least 100 million degrees Celsius.

2. The ions have to be confined together in close proximity to allow them to fuse. A suitable ion density is $2 - 3 \times 10^{20}$ ions/m^3.

3. The ions must be held together in close proximity at high temperature long enough to avoid plasma cooling.

At higher densities, charged particles in the plasma moving at high speeds may give rise to bremsstrahlung—radiation given off by a charged particle (most often an electron) due to its acceleration caused by an electric field of another charged particle (most often a proton or an atomic nucleus). Bremsstrahlung could become so dominant that all the energy in the plasma may radiate away. Other radiation losses, including synchrotron radiation from charged particles orbiting about magnetic fields would be negligible. A fusion reactor, therefore, has to be operated at a temperature where the power gain from fusion would exceed the bremsstrahlung losses.

6.2.2 Debye length

A parameter that determines the electrostatic properties of a plasma is called the Debye length L_D [7]:

$$L_D \propto \sqrt{\frac{k_B T}{n_d}}. \tag{17}$$

It is a length scale over which electrons screen out electric fields in the plasma. In other words, it is the distance over which significant charge separation can occur and how far its electrostatic effect persists. For distances greater than the Debye length, the energy of the particles in the plasma balances the electrostatic potential energy.

Using $n_d = 10^{28}$ particles/m^3, the Debye length for a 10 keV plasma is of the order of 10 nm, and the number of particles in a volume of the plasma of one Debye length is about 10^4. For a more rarefied plasma, say $n_d = 10^{22}$ particles/m^3, $L_D = 10$ μm, and the number of particles in a volume of dimension of one Debye length is 10^7. In either of these two extreme cases, there are two basic properties: the physical size of the plasma is far larger than the Debye length, and there are many particles in a spherical volume of radius equal to one Debye length. They are these two properties that describe the hot thermonuclear fuel.

7. Plasma confinement

Just like a conventional power plant, a fusion power plant will use the energy released during fusion reaction to produce steam and then generate electricity by way of turbines and generators. But as noted in the above discussions, it is hard to harness the energy in a laboratory environment.

Each fusion reaction is characterized by a specific ignition temperature, which must be surpassed before the reaction can occur. In stars, which are made of plasma, fusion takes place because of immense gravitational forces and extreme temperatures. Trying to create similar conditions here on Earth has required fundamental advances in a number of fields, from quantum physics to materials science. Scientists and engineers have made enough progress over the past half century, especially since the 1990s, so that a fusion reactor able to generate more power than it takes to operate can be built. Supercomputing has helped enormously, allowing researchers to precisely model the behavior of plasma under different conditions.

One of the major requirements in the development of a fusion reactor is the actual realization of the ignition temperature of d-t reaction, which is 100 million degrees Celsius. Once all the conditions are realized, the challenge to contain and control the staggering levels of heat in the plasma is formidable. That is because the plasma must not only be heated to a temperature of at least 100 million degrees Celsius, but the energy must also be confined within the plasma without being carried to walls of the container for times long enough for the relatively infrequent fusion events to occur. Otherwise, the plasma will exchange energy with the walls, cool itself down, and melt the container.

Many techniques have been developed, but the two main experimental approaches that seem capable of doing this task are magnetic confinement and inertial confinement.

7.1 Magnetic confinement

This method uses strong magnetic fields to contain the hot plasma and prevent it from coming into contact with the reactor walls. The magnetic fields keep the plasma in perpetually looping paths because the electrical charges on the separated ions and electrons mean that they follow the magnetic field lines. As a consequence, the plasma does not touch the wall of the container.

There are several types of magnetic confinement system, but the approaches that have been developed to the point of being used in a reactor are tokamak and stellarator devices. Because of its versatility, tokamak is considered to be the most developed magnetic confinement system. Hence, it is the workhorse of fusion.

7.1.1 Tokamak

The tokamak, acronym for the Russian phrase *toroidál'naja kámera s magnitnymi katúškami* meaning toroidal chamber with magnetic coils, was designed in 1951 by Soviet physicists Andrei Sakharov and Igor Tamm [8]. It is a doughnut-shaped device in which the combination of two sets of magnetic coils, known as toroidal and poloidal field coils, creates a field in both vertical and horizontal directions. The magnetic fields hold and shape the charged particles of the plasma by forcing them to follow the magnetic field lines. They essentially create a "cage," a magnetic bottle, inside which the plasma is confined. A strong electric current is induced in the plasma using a central solenoid, and this induced current also contributes to the poloidal field.

7.1.2 Stellarators

Unlike tokamaks, stellarators [9, 10] do not require a toroidal current to be induced in the plasma. Instead, the plasma is confined and heated by means of helical magnetic field lines. They are produced by a series of coils which may themselves be helical in shape. As a result, plasma stability is increased compared with tokamaks. Since heating the plasma can be more easily controlled and monitored with stellarators, they have an intrinsic potential for steady-state, continuous operation. The disadvantage is that, due to their more complex shape, stellarators are much more complicated than tokamaks to design and build.

7.2 Inertial confinement

After the invention of laser in 1960 at Hughes Research Laboratory in California, researchers sought to heat the fusion fuels with a laser so suddenly that the plasma would not have time to escape before it was burned in the fusion reaction. It would be trapped by its own inertia, hence the name "inertial confinement," because it relies on the inertia of the implosion to bring nuclei close together. This approach to confinement was developed at Lawrence Livermore National Laboratory in California [11].

Within the context of inertial confinement, laser beams with an intensity of the order of $10^{14} - 10^{15}\,W/cm^2$ are fired on a solid pellet filled with a low-density mixture of deuterium and tritium. The energy of the laser vaporizes the pellet instantly producing a surrounding plasma environment for a short period of time. During the process, the density and temperature of the fuel attains a high enough value to ignite the fusion reaction.

The capability of present lasers does not allow the inertial confinement technique to obtain break-even conditions, simply because the efficiency for converting electrical energy into radiation is very low, about 1–10%. Consequently, alternative approaches are being explored to achieve the ignition temperature. One such approach involves using beams of charged particles instead of lasers.

8. Cold fusion

In 1989, researchers at University of Utah (USA) and University of Southampton (UK) claimed to have achieved fusion at room temperature in a simple tabletop experiment involving the electrolysis of heavy water (deuterium oxide) using palladium electrodes [12]. According to them, when electric current passed through the water, palladium catalyzed fusion by allowing deuterium atoms to get close enough for fusion to occur. Since other experimenters failed to replicate their claim, most of the scientific community no longer considers it a real phenomenon.

But in 2005, cold fusion got a major boost. Scientists at UCLA initiated fusion using a pyroelectric crystal [13]. They put the crystal into a small container filled with hydrogen, warmed the crystal to produce an electric field, and inserted a metal wire into the container to focus the charge. The focused electric field powerfully repelled the positively charged hydrogen nuclei, and in the rush away from the wire, the nuclei smashed into each other with enough force to fuse. The reaction took place at room temperature.

9. Fusion research

The aim of the controlled fusion research program is to achieve ignition, which occurs when enough fusion reactions take place for the process to become self-sustaining, with fresh fuel then being added to continue it. Once ignition is achieved, there is a net energy yield—about four times as much as with nuclear fission. As mentioned earlier, such conditions can occur when the temperature increases, causing the ions in the plasma to move faster and eventually reach speeds high enough to bring the ions close enough together. The nuclei can then fuse, causing a release of energy.

The plasma temperature needed for ignition is produced by external heating. Powerful methods were developed for this purpose. They are:

1. Heating by injection of neutral beams: In this method, neutralized particles with high kinetic energy, produced in an ion source, are injected into the plasma, whereby they transfer their energy to the plasma through collisions.

2. Heating by high-frequency radio or microwaves: When electromagnetic waves of appropriate frequency are beamed into the plasma, the plasma particles absorb energy from the field of the wave and transfer it to the other particles through collisions.

3. Heating with current: When an electric current is passed through the plasma, it generates heat in the plasma through its resistance. As the resistance decreases with increasing temperature, this method is only suitable for initial heating.

These methods produce temperatures of 100 million degrees Celsius in present-day fusion devices.

9.1 Research programs

Experiments with d-t fuel began in the early 1990s in the Tokamak Fusion Test Reactor in Princeton (USA) [14] and the Joint European Torus (JET) in Culham (UK) [15]. The world's first controlled release of fusion power using a 50–50 mix of tritium and deuterium with a fusion output of 16 MW from an input of 24 MW heating (Q-factor is 0.67) was achieved in 1991 by JET. The Q-factor is used to represent the ratio of the power produced in the fusion reaction to the power required to produce the fusion. It should not be confused with the Q-value of a reaction, which is the amount of energy released by that reaction. Obviously, Q-factor of 1 is breakeven. To achieve commercially viable fusion energy, the Q-factor must be much greater than one.

The 35-nation International Thermonuclear Experimental Reactor (ITER, "The Way" in Latin) project currently under construction in Cadarache, France, is the world's largest tokamak fusion reactor [16]. The goals of ITER are:

1. To operate at 500 MW (for at least 400 s continuously) with less than 50 MW of input power for a tenfold energy gain (Q-factor is 10).

2. Demonstrate the integrated operation of technologies for a fusion power plant and test technologies for heating, control, diagnostics, cryogenics, and remote maintenance.

3. Achieve a deuterium-tritium plasma in which the reaction is sustained through internal heating and stays confined within the plasma efficiently enough for the reaction to be sustained for a long duration.

4. Test tritium breeding because the world supply of tritium (used with deuterium to fuel the fusion reaction) is not sufficient to cover the needs of future power plants.

5. Demonstrate the safety characteristics of a fusion device, particularly the control of the plasma and fusion reactions with negligible consequences to the environment.

Launched in 2006, the project has been beset with technical delays, labyrinthine decision-making, and cost estimates that have soared. The reactor is now expected to be completed and become operational by 2030.

According to ITER Newsletter [16], "When completed, the plasma circulating in the core of the reactor will be 150 million degrees Celsius, or about 10 times hotter than the Sun. The massive superconducting magnets surrounding the core will be cooled to −270 degrees, as cold as the depths of space. So many of the technologies involved are really at the cutting edge."

There is a considerable amount of research into many other fusion projects at various stages of development, but ITER is the largest, with 10 times more plasma capacity than any other reactor. Although China is a participating country in the ITER project, the Chinese are nevertheless building a tokamak reactor by themselves. Known as the Experimental Advanced Superconducting Tokamak (EAST), they managed to heat hydrogen gas to a temperature of about 50 million degrees Celsius [17].

Based on the information, technologies, and experience provided by ITER, physicists and engineers at the Culham Laboratory in Oxfordshire (UK) are working to develop a Demonstration Power Station (DEMO) which, if successful in terms of systems and performance, could be used as the commercial prototype, creating a fast track to fusion power. In collaboration with the Princeton Plasma Physics Laboratory, South Korea is also developing a tokamak fusion reactor named Korean Demonstration Fusion Power Plant (K-DEMO) [18]. Both EAST and K-DEMO are due for completion by year 2030.

Under an Italian-Russian agreement, Italy's National Agency for New Technologies, Energy and Sustainable Economic Development is developing a small tokamak reactor by the name of Ignitor [19]. The reactor is based on the Alcator machine at MIT [20] which pioneered the high magnetic field approach to plasma magnetic confinement. The scientists of the project believe that unlike the larger ITER reactor, Ignitor could be ready to begin operations within a few years.

By using magnetic fields that are twice as strong as those planned for ITER, two spin-off companies, one in the USA and the other in the UK, hope to create a sustainable fusion reaction in a machine as small as 1/70th the size of ITER. They also believe, according to the August 2018 issue of Physics Today, that their reactor will be able to produce more energy than they consume. It is expected to be operational before ITER, possibly by the mid-2020s.

The Germans are working on a non-tokamak fusion reactor called Wendelstein 7-X [21]. In a test run, they produced helium plasma that lasted for one-quarter of a second and achieved a temperature of 80 million degrees Celsius. The Germans believe that their stellarator design, similar in principal to the tokamaks, will provide an inherently more stable environment for plasma and a more promising route for nuclear fusion research in general.

Another stellarator, TJ-II, designed in collaboration with Oak Ridge National Laboratory (USA), is in operation in Madrid, Spain [22]. This flagship project of the National Fusion Laboratory of Spain is a flexible, medium-size stellarator—the second largest operational stellarator in Europe, after Wendelstein 7-X.

In 2014, scientists and engineers at the American aerospace conglomerate Lockheed Martin claimed to have made a major technological breakthrough in the development of a fusion reactor [23]. They are cautiously optimistic that an operational reactor with enough energy output to power a small city, yet small enough to fit on the back of a truck, can be built before the end of this decade. However, because of the absence of further details on how their reactor works, some scientists are skeptical about the claim.

According to MIT Technology Review [24], while ongoing research centered on large tokamaks may take decades before a commercially feasible fusion reactor is built, several privately funded companies and small university-based research groups pursuing novel fusion reactor designs have delivered promising results that could shorten the timeline for producing a prototype machine from decades to several years. On the other hand, scientists of the mega-projects believe that fusion power could become a reality more quickly if the present international funding for fusion research was increased.

There have also been significant developments in research into inertial confinement fusion (ICF). Research on ICF in the USA is going on at the National Ignition Facility [25] at the Lawrence Livermore National Laboratory in California and Sandia Laboratories in New Mexico. At Sandia, an entirely different method of ICF called the Z-pinch [26], which does not use laser at all, is being investigated. Instead, it uses a strong electrical current in a plasma to generate X-rays, which compresses a tiny d-t fuel cylinder. The other notable research activity on ICF is the Laser Megajoule project in Bordeaux, France [27]. All three projects are designed to deliver, in a few billionths of a second, nearly 2 million Joules of energy to targets measuring a few millimeters in size. The main purpose of these projects is, however, to support research for nuclear weapons programs.

Thus far, none of the ICF facilities have achieved scientific breakeven, which is a gain of unity. However, for making fusion energy viable in commercial power plants, the gain has to be much greater than breakeven. Since lasers are very inefficient machines, gains of at least 100 are needed for a plant to produce net power output. To that end, researchers at Lawrence Livermore National Laboratory are exploring other approaches to developing ICF as a source for energy.

10. Advantages/disadvantages of fusion reactors

There are many advantages of fusion reactors:

1. They will produce at least five times more energy than the amount of energy it will need to heat the fusing nuclei to the desired temperature. Furthermore, it is estimated that to run a 1000 MW power plant for a year, a fusion reactor will require about 3000 m^3 of water (source of deuterium) and 10 tonnes of lithium ore, while the current fission reactors consume 25–30 tonnes of enriched uranium. Clearly, gram for gram, fusion reactor wins the energy race by a wide margin.

2. Fusion fuels are widely available and nearly inexhaustible. Deuterium can be distilled from all forms of water, while reserves of lithium, both terrestrial and sea-based, which would be used to produce tritium, would fulfill needs of fusion reactors for millions of years.

3. Unlike fission, fusion will have a low burden of radioactive waste. They will not produce high-level nuclear wastes like their fission counterparts, so disposal will be less of a problem. Fusions by-product is helium—an inert, nontoxic, and nonradioactive gas used to inflate childrens' balloons. Besides, there will be no fissile material that could be diverted by terrorists to build "dirty bombs." Moreover, a fusion power station would not require the transport of hazardous radioactive materials.

4. Fusion reactors are inherently incapable of a runaway reaction that could result in a core meltdown, the most serious calamity possible in a fission reactor. This is because there is no critical mass required for fusion. Besides, fusion reactors work like a gas burner; once the fuel supply is shut off, the reaction stops. There will, therefore, be no off-site radiation-related deaths, even from a severe accident.

5. Despite being technically nonrenewable, fusion has many of the benefits of renewable energy sources, such as being a long-term source of energy emitting no greenhouse gases. Besides, because it is not dependent on weather, fusion could provide uninterrupted power delivery, unlike solar and wind power.

Although fusion does not generate long-lived radioactive products and the unburned gases can be treated on site, there are nevertheless few concerns related to the radioactivity induced by the high-energy neutrons ($\sim$14 MeV) that are produced during the d-t reaction. They are:

1. Some radioactive wastes will be produced due to neutron activation of lithium to produce tritium inside the reactor, but their inventory will be much less than those from fission, and they will be short-lived. Nonetheless, if accidentally released in the air or water, tritium will remain radioactive for a period equal to at least 10 half-lives or 120 years.

2. The neutrons will irradiate the surrounding structures giving rise to radioactive nuclides, which ultimately have to be disposed of in some waste facility. But their stock will be considerably lower than that from actinides used in fission-based reactors.

3. Since most of the energy in the d-t reaction is carried away by the neutrons, this could lead to neutron leakage that could be significantly higher than uranium reactors. More neutron leakage means more shielding and improved protection for workers at the power plant.

11. Fusion torch

A fascinating application for the abundant energy that fusion may provide is the fusion torch, a star-hot flame or high-temperature plasma into which all waste materials—whether liquid sewage or solid industrial refuse—could be dumped [28]. In the high-temperature environment, the materials would be reduced to their constituent atoms and separated by a mass-spectrograph-type device into various bins ranging from hydrogen to uranium. Thus, a single fusion plant could, in principle, not only dispose of thousands of tons of solid wastes per day but also convert them into a few reusable and saleable elements, thereby closing the cycle from use to reuse.

12. Conclusion

Projection for the demand of energy depends on the growth of population, because the more people there is, the more energy will be used. The current world population of 7 billion is expected to reach 11 billion in 2100. This means if we want to maintain a better or at least the same standard of living, global consumption of energy could double, or perhaps triple, by the end of this century.

With the incorporation of improved safety features and new generation of reactors, nuclear fission will probably continue to make a major contribution to electricity generation. However, its growth could be curtailed by issues of public and political acceptability. Supplies from some renewable sources of energy, such as solar or wind, are not guaranteed either, because they are reliant on weather conditions. Technological challenges for other sources, ocean thermal energy and hydrokinetic energy from rivers, for example, have not yet been fully developed. So, for future energy security, the answer is nuclear fusion.

Advocates acknowledge that fusion technology is likely many decades away. The reason, these systems are intrinsically large, so large that we cannot test the physics and technology of fusion on a lab bench and then mass-produce fusion reactors. Consequently, these large, first-of-a-kind facilities take time to construct.

Despite the enormity of the projects, we have succeeded in creating a short-lived artificial Sun on Earth via experimental fusion reactors. Once commercial fusion reactors become a reality, there will be a paradigm-shifting development in the global energy mix. In particular, our dependence on the rapidly depleting supply of fossil fuels and uranium will be drastically reduced. More importantly, fusion power can easily secure our planets future, given the abundance of fuels and near-limitless energy produced from fusion reactions. Additionally, with no risk for proliferation and minimum radioactive waste generated, nuclear fusion would offer a clean, relatively safe, zero greenhouse gas-emitting, and long-term source of energy, with the potential to produce at least 20–25% of the world's electricity by 2100.

To conclude, nuclear fusion energy may not have the magic wand that would solve our energy problem. Nevertheless, it has the potential to be an attractive energy source that can be deployed as major pressures rise on existing energy supply options. Also, it would go a long way in slowing down, if not mitigating, the unrelenting climb of the temperature of our planet.

Author details

Quamrul Haider
Department of Physics and Engineering Physics, Fordham University, Bronx, New York, USA

*Address all correspondence to: haider@fordham.edu

IntechOpen

© 2019 The Author(s). Licensee IntechOpen. This chapter is distributed under the terms of the Creative Commons Attribution License (http://creativecommons.org/licenses/by/3.0), which permits unrestricted use, distribution, and reproduction in any medium, provided the original work is properly cited.

References

[1] Meitner L, Frisch OR. Disintegration of uranium by neutrons: A new type of nuclear reaction. Nature. 1939;**143**:239

[2] Hooshyar MA, Reichstein I, Malik FB. Nuclear Fission and Cluster Radioactivity: An Energy-Density Functional Approach. New York: Springer; 2005

[3] World Nuclear Association. Physics of Uranium and Nuclear Energy. Available from: http://www.world-nuclear.org/information-library/nuclear-fuel-cycle/introduction/physics-of-nuclear-energy.aspx

[4] Bethe HA. Energy production in stars. Physical Review. 1939;**55**:434

[5] World Nuclear Association. Nuclear Fusion Power. Available from: http://www.world-nuclear.org/information-library/current-and-future-generation/nuclear-fusion-power.aspx

[6] Lawson JD. Some criteria for power producing themonuclear reactor. Proceedings of the Physical Society. 1957;**70**:6

[7] Know plasma, know 99.999% of the universe. Available from: http://www.plasma-universe.com

[8] Smirnov VP. Tokamak foundation in USSR/Russia 1950–1990. Nuclear Fusion. 2010;**50**:014003

[9] Wobig H. Theory of advanced stellarators. Plasma Physics and Controlled Fusion. 1999;**41**:A159

[10] Todd TN. MHD control and ECCD in Compass–D. Plasma Physics and Controlled Fusion. 1993;**35B**:231

[11] Nakai S, Takabe H. Principles of inertial confinement fusion – physics of implosion and the concept of inetial fusion energy. Reports on Progress in Physics. 1996;**59**:1071

[12] Fleischman M, Pons S. Electrochemically induced nuclear fusion of deuterium. Journal of Electroanalytical Chemistry. 1988; **261**:301

[13] Naranjo B, Gimzewski JK, Putterman S. Observation of nuclear fusion driven by a pyroelectric crystal. Nature. 2005;**434**:1115

[14] Princeton tokamak heats up the race for fusion power. Popular Science. 1978; **150**:69

[15] Culham Centre for Fusion Energy. Available from: http://www.ccfe.ac.uk/JET.aspx

[16] Motojima O. The ITER Project construction status. Nuclear Fusion. 2015;**55**:104023. Available from: https://www.iter.org/whatsnew

[17] China builds an "artificial sun" that is 6 times hotter than our "natural sun". Available from: https://www.techworm.net/2018/11/china-artificial-sun-six-times-hotter-natural-sun.html

[18] Kim K et al. A preliminay conceptual design study for Korean fusion DEMO reactor. Fusion Engineering and Design. 2013;**88**:488

[19] Italy and Russia revive Ignitor. Available from: https://www.iter.org/newsline/131/169

[20] Greenwald M et al. 20 years of research on the Alcator C-Mod tokamaka. Physics of Plasmas. 2014;**21**: 11052

[21] Beidler CD et al. Plasma Physics and Controlled Nuclear Fusion Research. In: Proceedings of the 9th International

Conference; Montreal. Vol. III. Vienna: IAEA; 1982. p. 129

[22] Navarro AP et al. Engineering aspects and present status of the Spanish Stellaator TJ-II. In: Fusion Engineering 14th IEE/NPSS Symposium. Vol. 1. 1997. p. 257

[23] Skunk Works reveals compact fusion reactor details. In: AviationWeek and Space Technology. 2014

[24] MIT Technology Rreview. Available from: https://www.technologyreview.com/magazine/2015/09

[25] NIF achieves record double fusion yield. Available from: https://wci.llnl.gov/news/inertial-confinement-fusion

[26] Inertial confinement fusion. Available from: https://www.sandia.gov/Pulsed-Power/res-areas/inertial_confinement

[27] Le Laser Mégajoule: Available from: http://www-lmj.cea.fr/index-en.htm

[28] Eastlund BJ, GoughWC, Sabri ZA, Rose DJ. In defense of fusion torch. Physics Today. 1972;**25**:10

Chapter 2

Fusion Reaction of Weakly Bound Nuclei

Fouad A. Majeed, Yousif A. Abdul-Hussien
and Fatima M. Hussian

Abstract

Semiclassical and full quantum mechanical approaches are used to study the effect of channel coupling on the calculations of the total fusion reaction cross section σ_{fus} and the fusion barrier distribution D_{fus} for the systems $^{6}Li + ^{64}Ni$, $^{11}B + ^{159}Tb$, and $^{12}C + ^{9}Be$. The semiclassical approach used in the present work is based on the method of the Alder and Winther for Coulomb excitation. Full quantum coupled-channel calculations are carried out using CCFULL code with all order coupling in comparison with our semiclassical approach. The semiclassical calculations agree remarkably with the full quantum mechanical calculations. The results obtained from our semiclassical calculations are compared with the available experimental data and with full quantum coupled-channel calculations. The comparison with the experimental data shows that the full quantum coupled channels are better than semiclassical approach in the calculations of the total fusion cross section σ_{fus} and the fusion barrier distribution D_{fus}.

Keywords: fusion reaction, breakup channel, weakly bound nuclei, fusion barrier

1. Introduction

In recent years, big theoretical and experimental efforts had been dedicated to expertise the effect of breakup of weakly bound nuclei on fusion cross sections [1, 2]. This subject attracts special interests for researchers and scholars, because the fusion of very weakly bound nuclei and exotic radioactive nuclei is reactions that have special interests in astrophysics which play a very vital role in formation of superheavy isotopes for future applications [3–8]. Since the breakup is very important to be considered in the fusion reaction of weakly bound nuclei, the following should be considered: the elastic breakup (EBU) in which neither of the fragments is captured by the target; incomplete fusion reaction (ICF), which happens when one of the fragments, is captured by the target; and complete fusion following BU (CFBU), which happens in all breakup fragments that are captured by the target, is called the sequential complete fusion (SCF). Therefore, the total breakup cross section is the sum of three contributions: EBU, ICF, and CFBU, whereas the sum of complete fusion (including two body fusions and CFBU) and incomplete fusion is called total fusion (TF) [1, 8–10]. Fusion reactions with high-intensity stable beams which have a significant breakup probability are good references for testing the models of breakup and fusion currently being developed. The light nuclei such as ^{6}Li breakup into $^{4}He+^{2}H$, with separation energy $S\alpha$ = 1.48 MeV; 11B breakup into

$^{4}He+^{7}Li$ with separation energy $S\alpha$ = 8.664 MeV and ^{12}C breakup into three α particles induced by neutrons or protons by ^{12}C (p, p′) 3α [3, 11, 12]. The breakup channel is described by the continuum discretized coupled-channel (CDCC) method. The continuum that describes the breakup channel is discretized into bins [13, 14]. To study the coupled-channel problem, this requires a profound truncation of the continuum into discrete bin of energy into equally spaced states. The CDCC method is totally based on this concept. Surrey group extended the discretization procedure discussed in [14] for the deuteron case to study the breakup and fusion reactions of systems involving weakly bound nuclei [15, 16]. Recently, Majeed and Abdul-Hussien [17] utilized the semiclassical approach based on the theory of Alder and Winther. They carried out their calculations to investigate the role of the breakup channel on the fusion cross section σ_{fus} and fusion barrier distribution D_{fus} for $^{6,8}H$ halo [17]. Semiclassical coupled-channel calculations in heavy-ion fusion reactions for the systems ^{40}Ar + ^{110}Pd and ^{132}Sn + ^{48}Ca were carried out by Majeed et al. [18]. They argued that including the channel coupling between the elastic channel and the continuum enhances the fusion reaction cross section σ_{fus} and the fusion barrier distribution D_{fus} calculations quite well below and above the Coulomb barrier for medium and heavy systems. This study aims to employ a semiclassical approach by adopting Alder and Winther (AW) [19] theory originally used to treat the Coulomb excitation of nuclei. The semiclassical approach has been implemented and coded using FORTRAN programming codenamed (SCF) which is written and developed by Canto et al. [20]. The fusion cross section σ_{fus} and fusion barrier distribution D_{fus} are calculated here utilizing the semiclassical approach. The results from the present study are compared with the quantum mechanical calculations using the FORTRAN code (CC) [21] and with the experimental data for the three systems ^{6}Li + ^{64}Ni, ^{11}B + ^{159}Tb, and ^{12}C + ^{9}Be.

2. The semiclassical theory

2.1 The single-channel description

The semiclassical theory is used to estimate the fusion cross section in the one-dimensional potential model which assumes that one can describe the degree of freedom only of the relative motion between the colliding heavy ions [19, 20]. The semiclassical theory deals with the Schrödinger equation assuming independent energy and angular momentum and the potential energy for the radial part of the relative motion through quantum tunneling:

$$\left[-\hbar^2\nabla^2/2\mu + V(r) - E\right]\Psi(\mathrm{r}) = 0 \tag{1}$$

where μ is the reduced mass of the system and $V(r)$ is the total potential energy of the system. Semiclassical reaction amplitudes can be evaluated as a function of time, assuming the particle trajectory to be determined by classical dynamics, including Coulomb and real nuclear and centrifugal potentials, which can be written in the form

$$V(r) = V_C(r) + V_N(r) + V_l(r) \tag{2}$$

In coupled-channel effects on the elastic channel, the imaginary part should be added to the nuclear potential, represented by complex potential as

$$V_N(r) = U_N(r) - iW(r) \tag{3}$$

The method can be extended to describe interference of different l waves due to strong nuclear attraction and absorption caused by the imaginary nuclear potential [6, 20, 22]. When the two nuclei come across the potential barrier into the inner region, the fusion occurs according to the semiclassical theory, and the penetrability probability below barrier can be evaluated using WKB approximation [5, 6, 19, 23, 24]:

$$P_{fus}^{WKB}(l,E) = \frac{1}{1+\exp\left[2\int_{r_b^{(l)}}^{r_a^{(l)}} \kappa_l(r)dr\right]} \tag{4}$$

Then, the latter can be rewritten as follows:

$$P_{fus}^{WKB}(l,E) = \frac{1}{1+\exp\left[\frac{2\pi}{\hbar\Omega_l}(V_b(l)-E)\right]} \tag{5}$$

where $\kappa_l(r)$ is the local wave number and $r_b^{(l)}$ and $r_a^{(l)}$ are the inner and outer classical turning points at the fusion barrier potential. If one approximates the fusion barrier by a parabolic function, then the penetrability probability above barrier is given by the Hill-Wheeler formula [3, 19]:

$$P_{fus}^{WH}(l,E) = \frac{1}{1+\exp\left[\frac{2\pi}{\hbar\Omega_l}(E-V_b(l))\right]} \tag{6}$$

where $V_b(l)$ and Ω_l are the height and the curvature parameter of the fusion barrier for the partial wave, respectively, and E is the bombarding energy. Ignoring the l dependence of ω and of the barrier position R_b and assuming that the l dependence of $V_b(l)$ is given only by the difference of the centrifugal potential energy, one can obtain Wong's formula which is given in Section 5. The fusion cross sections can be estimated by the one-dimensional WKB approximation by the following relations [21, 24]:

$$\sigma_{fus}(E) = \frac{\pi}{\kappa^2}\sum(2l+1)P_{fus}^{WKB}(l,E) \tag{7}$$

$$P_{fus}^{\gamma}(l,E) = \frac{4k}{E}\int dr\left|u_{\gamma l}(k_\gamma,r)\right|^2 W_{fus}^{\gamma}(r) \tag{8}$$

where $u_{\gamma l}(k_\gamma, r)$ represents the radial wave function for the partial wave l in channel γ and $W_{fus}^{\gamma}(r)$ is the absolute value of the imaginary part of the potential associated to fusion in that channel.

The complete fusion cross section in heavy ions evaluated using semiclassical theory is based on the classical trajectory approximation r. And the relevant intrinsic degrees of freedom of the projectile, represented by ξ with applying the continuum discretized coupled-channel (CDCC) approximation of Alder and Winther (AW) theory [16], have been proposed [17]. The projectile Hamiltonian is then given by

$$h = h_0(\xi) + V(\xi, r) \tag{9}$$

where $\mathrm{h_0(\xi)}$ is the intrinsic Hamiltonian and $\mathrm{V(\xi,r) = V_N(\xi,r) + V_C(\xi,r)}$, $\mathrm{V(\xi,r)}$ is the interaction between the projectile and target nuclei. The Rutherford trajectory depends on the collision energy, E, and the angular momentum, l. In this case the trajectory is the solution of the classical motion equations with the potential

$V(r) = \langle\Psi_0|V(r,\xi)|\Psi_0\rangle$, where Ψ_0 is the ground state (g.s.) of the projectile. In this way, the interaction becomes time-dependent in the ξ-space $V_l(\xi.t) = V\left(r_{l(t)}, \xi\right)$, and the eigenstates of the intrinsic Hamiltonian $|\Psi_\gamma\rangle$ satisfy the Schrödinger equation [25, 26]:

$$h|\Psi_\gamma\rangle = \varepsilon|\Psi_\gamma\rangle \tag{10}$$

After expanding the wave function in the basis of intrinsic eigenstates

$$\Psi(\xi,t) = \sum a_\gamma(l,t)\Psi_\gamma(\xi)e^{-i\varepsilon_\gamma t/\hbar} \tag{11}$$

and inserting Eq. (11) into the Schrödinger equation for $\Psi(\xi,t)$, the AW equations can be obtained:

$$i\hbar\, \dot{a}_\gamma(l,t) = \sum_\epsilon \langle\Psi_\gamma|V(\xi,t)|\Psi_\gamma\rangle e^{i(\varepsilon_\gamma-\varepsilon_\epsilon)t/\hbar}\gamma\epsilon(l,t) \tag{12}$$

These equations should be solved with initial conditions $a_\gamma(l,t\to-\infty) = \delta_{\gamma 0}$ which mean that before the collision $(t\to-\infty)$, the projectile was in its ground state. The final population of the channel γ in a collision with angular momentum l is $P^\gamma_{fus}(l,E) = \left|a_\gamma(l,t\to-\infty)\right|^2$. Eq. (8) gives the general expression for the fusion cross section in multichannel scattering [26].

2.2 The coupled-channel description

The variables employed to describe the collision are the projectile-target separation vector $\vec{r}$ and the relevant intrinsic degrees of freedom of the projectile ξ. For simplicity, we neglect the internal structure of the target. The Hamiltonian is then given by [27]

$$H = H_0(\xi) + V\left(\vec{r},\xi\right) \tag{13}$$

where $H_0(\xi)$ is the intrinsic Hamiltonian of the projectile and $V(\vec{r},\xi)$ represents the projectile-target interaction. Since the main purpose of the present work is to test the semiclassical model in calculations of sub-barrier fusion, the nuclear coupling is neglected. Furthermore, for the present theory-theory comparison, only the Coulomb dipole term is taken into account. The eigenvectors of $H_0(\xi)$ are given by the equation [27]

$$H_0|\varphi_\beta\rangle = \varepsilon_\beta|\varphi_\beta\rangle \tag{14}$$

where ε_β is energy of internal motion. The AW method is implemented in two steps. First, one employs classical mechanics for the time evolution of the variable $\vec{r}$. The ensuing trajectory depends on the collision energy, E, and the angular momentum, ℓ. In its original version, an energy symmetrized Rutherford trajectory $\vec{r_\ell}(t)$ was used. In our case, the trajectory is the solution of the classical equations of motion with the potential $V(\vec{r}) = \langle\varphi_0|V(\vec{r},\xi)\varphi_0\rangle$, where $|\varphi_0\rangle$ is the ground state of the projectile. In this way, the coupling interaction becomes a time-dependent interaction in the ξ-space, $V_\ell(\xi,t) \equiv V(\vec{r_\ell}(t),\xi)$. The second step consists in treating the dynamics in the intrinsic space as a time-dependent quantum mechanics problem. Expanding the wave function in the basis of intrinsic eigenstates [19]

$$\psi(\xi,t) = \sum_{\beta} a_{\beta}(\ell,t)\varphi_{\beta}(\xi)e^{-i\varepsilon_{\beta}t/\hbar} \tag{15}$$

and inserting this expansion into the Schrödinger equation for $\psi(\xi,t)$, one obtains the AW equations [27]

$$i\hbar\dot{a}_{\beta}(\ell,t) = \sum_{\alpha} a_{\gamma}(\ell,t)\langle\varphi_{\beta}|V_{\ell}(\xi,t)|\varphi_{\beta}\rangle e^{-i(\varepsilon_{\beta}-\varepsilon_{\gamma})t/\hbar} \tag{16}$$

These equations are solved with the initial conditions $a_{\beta}(\ell,t\to-\infty) = \delta_{\beta 0}$, which means that before the collision ($t\to-\infty$), the projectile was in its ground state. The final population of channel β in a collision with angular momentum ℓ is $P_{\ell}^{(\beta)} = |a_{\beta}(\ell,t\to+\infty)|^2$ and the angle-integrated cross section is [19]

$$\sigma_{\beta} = \frac{\pi}{k^2}\sum_{\ell}(2\ell+1)P_{\ell}^{(\beta)} \tag{17}$$

To extend this method to fusion reactions, we start with the quantum mechanical calculation of the fusion cross section in a coupled-channel problem. For simplicity, we assume that all channels are bound and have zero spin. The fusion cross section is a sum of contributions from each channel. Carrying out partial-wave expansions, we get [28]

$$\sigma_F = \sum_{\beta}\left[\frac{\pi}{k^2}\sum_{\ell}(2\ell+1)P_{\ell}^{F}(\beta)\right] \tag{18}$$

with

$$P_{\ell}^{F}(\beta) = \frac{4k}{E}\int W_{\beta}^{F}(r)|u_{\beta\ell}(k_{\beta},r)|^2 dr \tag{19}$$

Above, $u_{\beta\ell}(k_{\beta},r)$ represents the radial wave function for the ℓth partial wave in channel β, and W_{β}^{F} is the absolute value of the imaginary part of the optical potential associated to fusion in that channel.

To use the AW method to evaluate the fusion cross section, we make the approximation [27]

$$P_{\ell}^{F}(\beta) \simeq \overline{P}_{\ell}^{(\beta)}T_{\ell}^{(\beta)}(E_{\beta}) \tag{20}$$

where $\overline{P}_{\ell}^{(\beta)}$ is the probability that the system is in channel β at the point of closest approach on the classical trajectory and $T_{\ell}^{(\beta)}(E_{\beta})$ is the probability that a particle with energy $E_{\beta} = E - \varepsilon_{\beta}$ and reduced mass $\mu = M_P M_T/(M_P + M_T)$, where M_P, M_T are the masses of the projectile and target, respectively, tunnels through the potential barrier in channel β [19].

We now proceed to study the CF cross sections in reactions induced by weakly bound projectiles. For simplicity, we assume that the g.s. is the only bound state of the projectile and that the breakup process produces only two fragments, F_1 and F_2. In this way, the labels $\beta = 0$ and $\beta \neq 0$ correspond, respectively, to the g.s. and the breakup states represented by two unbound fragments. Neglecting any sequential contribution, the CF can only arise from the elastic channel. In this way, the cross section σ_{CF} can be obtained from Eq. (20), dropping contributions from $\beta \neq 0$. That is [19],

$$\sigma_{CF} = \frac{\pi}{k^2} \sum_{\ell} (2\ell + 1) P_{\ell}^{Surv} T_{\ell}^{(0)}(E) \tag{21}$$

where

$$P_{\ell}^{Surv} \equiv \overline{P}_{\ell}^{(0)} = |a_0(\ell, t_{ca})|^2 \tag{22}$$

is usually called survival (to breakup) probability [19].

3. Fusion barrier distribution

Nuclear fusion is related to the transmission of the incident wave through a potential barrier, resulting from nuclear attraction plus Coulomb repulsion. However, the meaning of the fusion barrier depends on the description of the collision. Coupled-channel calculations include static barriers, corresponding to frozen densities of the projectile and the target. Its most dramatic consequence is the enhancement of the total fusion reaction cross section σ_{fus} at Coulomb barrier energies V_b, in some cases by several orders of magnitude. One of the possible ways to describe the effect of coupling channels is a division of the fusion barrier into several, the so-called fusion barrier distribution D_{fus} and given by [20, 29]

$$D_{fus}(E) = \frac{d^2F(E)}{dE^2} \tag{23}$$

where $F(E)$ is related to the total fusion reaction cross section through [29]

$$F(E) = E\sigma_{fus}(E) \tag{24}$$

The experimental determination of the fusion reaction barrier distribution has led to significant progress in the understanding of fusion reaction. This comes about because, as already mentioned, the fusion reaction barrier distribution gives information on the coupling channels appearing in the collision. However, we note from Eq. (24) that, since fusion reaction barrier distribution should be extracted from the values of the total fusion reaction cross section, it is the subject to experimental as well as numerical uncertainties [29–31]:

$$D_{fus}(E) \approx \frac{F(E + \Delta E) + F(E - \Delta E) - 2F(E)}{\Delta E^2} \tag{25}$$

where ΔE is the energy interval between measurements of the total fusion reaction cross section. From Eq. (25), one finds that the statistical error associated with the fusion reaction barrier distribution is approximately given by [30]

$$\delta D_{fus}^{stat}(E) \approx \frac{\sqrt{[\delta F(E + \Delta E)]^2 + [\delta F(E - \Delta E)]^2 + 4[\delta F(E)]^2}}{(\Delta E)^2} \tag{26}$$

where $\delta F(E)$ means the uncertainty in the measurement of the product of the energy by the total fusion reaction cross section at a given value of the collision energy. Then the uncertainties can approximately be written as [30]

$$\delta D_{fus}^{stat}(E) \approx \frac{\sqrt{6}\delta F(E)}{(\Delta E)^2} \tag{27}$$

4. Results and discussion

In this section, the theoretical calculations are obtained for total fusion reaction σ_{fus}, and the fusion barrier distribution D_{fus} using the semiclassical theory adopted the continuum discretized coupled channel (CDCC) to describe the effect of the breakup channel on fusion processes. The semiclassical calculations are carried out using the (SCF) code, while the full quantum mechanical calculations are performed by using the code (CC) for the systems $^{6}Li + ^{64}Ni$, $^{11}B + ^{159}Tb$ and $^{12}C + ^{9}Be$. The values of the height V_c, radius R_c, and curvature $\hbar\omega$ for the fusion barrier are displayed in **Table 1**.

4.1 The reaction $^{6}Li + ^{64}Ni$

The calculations of the fusion cross section σ_{fus} and fusion barrier distribution D_{fus} are presented in **Figure 1** panel (a) and panel (b), respectively, for the system $^{6}Li + ^{64}Ni$. The dashed blue and red curves represent the semiclassical and full quantum mechanical calculations without coupling, respectively. The solid blue and red curves are the calculations including the coupling effects for the semiclassical and full quantum mechanical calculations, respectively. Panel (a) shows the comparison between our semiclassical and full quantum mechanical calculations with the respective experimental data (solid circles).

The experimental data for this system are obtained from Ref. [32]. The real and imaginary Akyüz-Winther potential parameters obtained by using chi-square

System	V_c	R_c	$\hbar\omega$	Refs
$^{6}Li + ^{64}Ni$	12.41	9.1	3.9	[32]
$^{11}B + ^{159}Tb$	40.34	10.89	4.42	[33]
$^{12}C + ^{9}Be$	4.28	7.43	2.61	[34]

Table 1.
The fusion barrier parameters are height V_c (MeV), radius R_c (fm), and curvature $\hbar\omega$ (MeV).

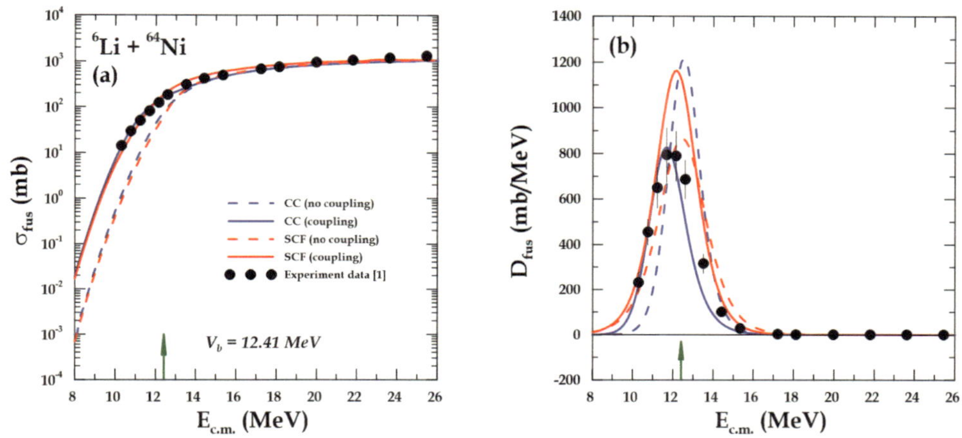

Figure 1.
The comparison of the coupled-channel calculations of semiclassical treatment (red curves) and full quantum mechanical (blue curves) with the experimental data of complete fusion (black-filled circles) [32] for $^{6}Li + ^{64}Ni$ system. Panel (a) refers to the total fusion reaction cross section σ_{fus} (mb), and panel (b) provides the fusion reaction barrier distribution D_{fus} (mb/MeV).

method are the strength $W_0 = 50\,MeV$, radius $r_i = 1.0\,fm$, and diffuseness $a_i = 0.25\,fm$, and for the real part, the depth is $V_0 = 35.0\,MeV$, radius is $r_0 = 1.1\,fm$, and diffuseness is $a_0 = 0.8\,fm$. The χ^2 values obtained for the total fusion cross section σ_{fus} are 1.5057 and 1.1286 in the case of no coupling for semi-classical and quantum mechanical calculations, respectively. The χ^2 values obtained for the case of coupling effects included are 0.2431 and 0.3115 for semiclassical and quantum mechanical calculations, respectively. The χ^2 values show clearly that semiclassical calculations including coupling effects are more consistent with the experimental data than full quantum mechanical including coupling effects. The χ^2 values obtained using single-channel calculations for the fusion reaction barrier distribution D_{fus} are 0.1823 and 1.1914 for semiclassical and quantum mechanical calculations, respectively. The χ^2 values obtained when coupled channels are included are 0.1827 and 0.1321 for semiclassical and quantum mechanical calculations, respectively; the fusion barrier distribution D_{fus} has been extracted from the experimental data using Wong fit model along with the three-point difference method. The comparison with the experimental data for D_{fus} shows that the quantum mechanical calculations are in better agreement than the semiclassical calculations including the coupling effects.

4.2 The reaction $^{11}B + {}^{159}Tb$

In similar analysis we compare our theoretical calculations of the fusion cross section σ_{fus} and fusion barrier distribution D_{fus} with the corresponding experimental data in panels (a) and (b) of **Figure 2**, respectively, for the system $^{11}B + {}^{159}Tb$. The experimental data for this system is obtained from Ref. [33]. The real and imaginary Akyüz-Winther potential parameters are obtained by using chi-square method: $V_0 = 126.1$ MeV, $r_0 = 1.2$ fm, and $a_0 = 0.5$ fm and $W_0 = 55.9$ MeV, $r_i = 0.986$ fm, and $a_i = 0.614$ fm. The χ^2 values 0.9473 and 0.2486 are obtained for σ_{fus} using semiclassical and quantum mechanical distribution calculations without including the coupling, respectively, while the χ^2 for σ_{fus} using the semiclassical and quantum mechanical distribution calculations including the

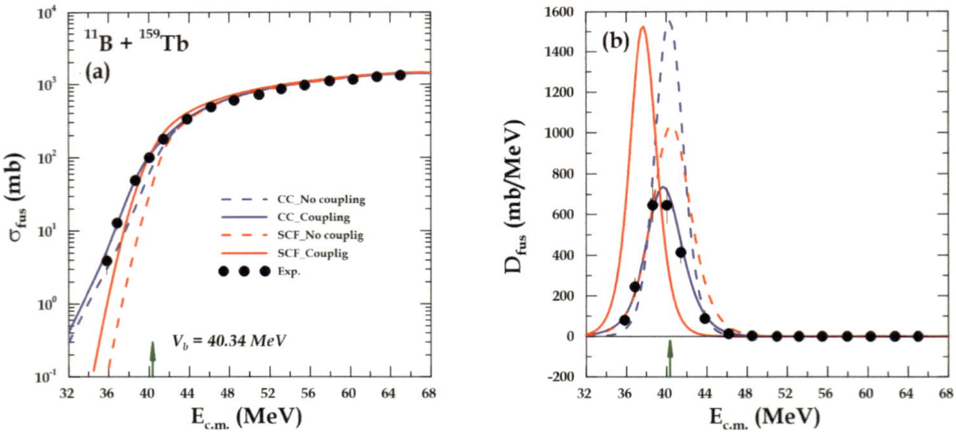

Figure 2.
The comparison of the coupled-channel calculations of semiclassical treatment (red curves) and full quantum mechanical (blue curves) with the experimental data of complete fusion (black-filled circles) [33] for $^{11}B + {}^{159}Tb$ system. Panel (a) shows the total fusion reaction cross section σ_{fus} (mb), and panel (b) gives the fusion reaction barrier distribution D_{fus} (mb/MeV).

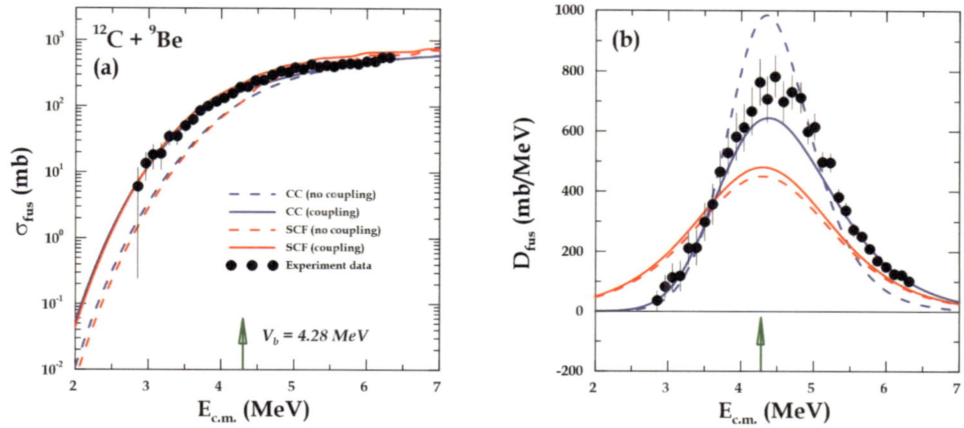

Figure 3.
The comparison of the coupled-channel calculations of semiclassical treatment (red curves) and full quantum mechanical (blue curves) with the experimental data of complete fusion (black-filled circles) [34] for $^{12}C + {}^{9}Be$ system. Panel (a) shows the total fusion reaction cross section σ_{fus} (mb), and panel (b) gives the fusion reaction barrier distribution D_{fus} (mb/MeV).

coupling effects are 0.2681 and 0.1657, respectively. The χ^2 for the fusion barrier distribution D_{fus} using the semiclassical and quantum calculations are 0.5828 and 1.2329 for no coupling and 4.5969 and 0.0616 including coupling effects, respectively. The χ^2 values for σ_{fus} and D_{fus} give clear evidence that the quantum mechanical calculations are in better agreement than the semiclassical calculations as compared with experimental data.

4.3 The reaction $^{12}C + {}^{9}Be$

Figure 3 (panels (a) and (b)) presents the comparison between our theoretical calculations for σ_{fus} and D_{fus} using both semiclassical and quantum mechanical calculations with the corresponding experimental data for the system $^{12}C + {}^{9}Be$. The experimental data for this system are obtained from Ref. [34]. The real and imaginary Akyüz-Winther potential parameters are obtained by using chi-square method: $V_0 = 40.3\,MeV$, $r_0 = 1.11\,fm$, $a_0 = 0.590\,fm$, $W_0 = 0\,MeV$, $r_i = 1.1\,fm$, and $a_i = 0.50\,fm$. The χ^2 values obtained from the comparison between the results and experimental data for σ_{fus} are 1.0633 and 1.1447 without coupling, and 0.4924 and 0.2072 with coupling, for semiclassical and quantum mechanical calculations, respectively. The obtained χ^2 values for D_{fus} using semiclassical and quantum mechanical calculations are 1.2383 and 0.6185 without coupling and 0.9875 and 0.1868 with account for coupling, respectively.

5. Conclusion

The semiclassical and quantum mechanical calculations for the total fusion reaction σ_{fus} and the fusion barrier distribution D_{fus} calculations below and around Coulomb barrier were discussed for the systems $^{6}Li + {}^{64}Ni$, $^{11}B + {}^{159}Tb$, and $^{12}C + {}^{9}Be$. We conclude that the breakup channel is very important to be taken into consideration to describe the total fusion reaction σ_{fus} and the fusion barrier distribution D_{fus} for reaction of light projectiles. The full quantum mechanical calculations are closer to the experimental data than the semiclassical calculations;

however, semiclassical ones can be considered a successful tool for studying fusion reaction of systems involving light projectiles.

Acknowledgements

The author FA Majeed gratefully acknowledges financial assistance from Conselho Nacional de Desenvolvimento Científico e Tecnológico (CNPq) (Brazil) and is especially indebted to the World Academy of Sciences for the advancement of science in developing countries (TWAS) (Italy) for a 1-year grant under the scheme (TWAS-CNPq exchange programs for postdoctoral researchers).

Author details

Fouad A. Majeed*, Yousif A. Abdul-Hussien and Fatima M. Hussian
Department of Physics, College of Education for Pure Sciences, University of Babylon, Babylon, Iraq

*Address all correspondence to: fouadattia@gmail.com

IntechOpen

© 2019 The Author(s). Licensee IntechOpen. This chapter is distributed under the terms of the Creative Commons Attribution License (http://creativecommons.org/licenses/by/3.0), which permits unrestricted use, distribution, and reproduction in any medium, provided the original work is properly cited.

References

[1] Gomes PRS, Padron I, Crema E, Capurro OA, Fernández Niello JO, et al. Comprehensive study of reaction mechanisms for the ^{9}Be+^{144}Sm system at near- and sub-barrier energies. Physical Review C. 2006;**73**:064606

[2] Gomes PRS, Rios JL, Borges JR, Otomar DR. Fusion, breakup and scattering of weakly bound nuclei at near barrier energies. The Open Nuclear & Particle Physics Journal. 2013;**6**:10-15

[3] Dasgupta M, Gomes PRS, Hinde DJ, Moraes SB, Anjos RM, Berriman AC, et al. Effect of breakup on the fusion of ^{6}Li, ^{7}Li, and ^{9}Be with heavy nuclei. Physical Review C. 2004;**70**:024606

[4] Takigawa N, Sagawa H. Interaction potential and fusion of a halo nucleus. Physics Letters B. 1991;**265**:23-28

[5] Hussein MS, Pato MP, Canto LF, Donangelo R. Near-barrier fusion of ^{11}Li with heavy spherical and deformed targets. Physical Review C. 1992;**46**: 377-379

[6] Dasso CH, Vitturi A. Does the presence of ^{11}Li breakup channels reduce the cross section for fusion processes? Physical Review C. 1994;**50**:R12-R14

[7] Canto LF, Donangelo R, Lotti P. Effect of Coulomb dipole polarizability of halo nuclei on their near-barrier fusion with heavy targets. Physical Review C. 1995;**52**:R2848-R2850

[8] Martí GV, Gomes PRS, Rodríguez MD, Fernández Niello JO. Fusion, reaction, and breakup cross sections of ^{9}Be on a light mass target. Physical Review C. 2005;**71**:027602

[9] Padron I, Gomes PRS, Anjos RM, Lubian J, Muri C, Alves JJ. Fusion of stable weakly bound nuclei with ^{27}Al and ^{64}Zn. Physical Review C. 2002;**66**: 044608

[10] Wang B, Zhao W, Gomes PRS, Zhao E, Zhou S. Systematic study of breakup effects on complete fusion at energies above the Coulomb barrier. Physical Review C. 2014;**90**:034612

[11] Wolfs FLH, White CA, Bryan DC, Freeman CG, Herrick DM, et al. Breakup of 82 MeV ^{11}B. Physical Review C. 1994;**49**:2538-2548

[12] Bogatin VI, Novak Z, Ostroumov VI. Breakup of 12C into three alpha particles accompanying inelastic scattering of 80-MeV π^{+} mesons. Soviet Physics - JETP. 1963;**16**:1116-1121

[13] Matsumoto T, Hiyama E, Ogata K, Iseri Y, Kamimura M, Chiba S, et al. Continuum-discretized coupled-channels method for four-body nuclear breakup in ^{6}He+^{12}C scattering. Physical Review C. 2004;**0616061**(R):70

[14] Austern N, Iseri Y, Kamimura M, Kawai M, Rawitscher G, Yahiro M. Continuum-discretized coupled-channels calculations for three-body models of deuteron-nucleus reactions. Physics Reports. 1987;**154**:125-204

[15] Nunes FM, Thompson IJ. Nuclear interference effects in ^{8}B sub-Coulomb breakup. Physical Review C. 1998;**57**: R2818-R2820

[16] Tostevin JA, Nunes FM, Thompson IJ. Calculations of three-body observable in ^{8}B breakup. Physical Review C. 2001;**63**:024617

[17] Majeed FA, Abdul-Hussien YA. Semiclassical treatment of fusion and breakup processes of ^{6,8}He halo nuclei. Journal of Theoretical and Applied Physics. 2016;**107**:1-6

[18] Majeed FA, Hamodi RSh, Hussian FM. Journal of Computational and Theoretical Nanoscience. 2017;**14**:2242-2247

[19] Alder K, Winther A. Electromagnetic Excitations. Amsterdam: North-Holland; 1975

[20] Canto LF, Donangelo R, Marta HD. Physical Review C. 2006;**73**:034608 SCF code unpublished

[21] Canto LF, Gomes PRS, Donangelo R, Hussein MS. Fusion and breakup of weakly bound nuclei. Physics Reports. 2006;**424**:1-111

[22] Gomes PRS, Rodríguez MD, Martí GV, Padron I, Chamon LC. Effect of the breakup on the fusion and elastic scattering of weakly bound projectiles on ^{64}Zn. Physical Review C. 2005;**71**: 034608

[23] Canto LF, Gomes PRS, Donangelo R, Hussein MS. Recent developments in fusion and direct reactions with weakly bound nuclei. Physics Reports. 2015; **596**:1-86

[24] Cárdenas WHZ, Canto LF, Donangelo R, Hussein MS, Lubian J, Romanelli A. Approximations in fusion and breakup reactions induced by radioactive beams. Nuclear Physics A. 2002;**703**:633-648

[25] Hussein MS, Pato MP, Canto LF, Donangelo R. Real part of the polarization potential for ^{11}Li-induced fusion reactions. Physical Review C. 1993;**47**:2398-2400

[26] Ibraheem AA, Bonaccors A. Coulomb breakup effects on the optical potentials of weakly bound nuclei. Nuclear Physics A. 2005;**748**:414-432

[27] Liang JF, Signorini C. Fusion induced by radioactive ion beams. International Journal of Modern Physics E: Nuclear Physics. 2005;**14**:1121

[28] Abramowitz M, Stegun I. Handbook of Mathematical Functions. New York: Dover; 1972

[29] Rowley N, Satchler GR, Stelson PH. On the "distribution of barriers" interpretation of heavy-ion fusion. Physics Letters B. 1991;**254**:25

[30] Bertulani CA, Canto LF. Semiclassical calculation of Coulomb break-up of weakly-bound nuclei. Nuclear Physics A. 1992;**539**:163

[31] Dasgupta M, Hinde DJ, Rowley N, Stefanini AM. Measuring barrier to fusion. Annual Review of Nuclear and Particle Science. 1998;**48**:401

[32] Shaikh MM, Roy S, Rajbanshi S, Pradhan MK, Mukherjee A, et al. Barrier distribution functions for the system ^{6}Li +^{64}Ni and the effect of channel coupling. Physical Review C. 2015;**91**: 034615

[33] Mukherjee M, Roy S, Pradhan MK, Sarkar MS, Basu P, et al. Influence of projectile α breakup threshold on complete fusion. Physics Letters B. 2006;**636**:91-95

[34] Cheung HC, High MD, Čujec B. Fusion and neutron transfer cross sections for ^{9}Be+^{12}C at sub coulomb energies. Nuclear Physics A. 1978;**296**: 333-348

Chapter 3

Fusion Neutronics Experiments for Thorium Assemblies

Rong Liu

Abstract

Thorium is a fertile element that can be applied in the conceptual blanket design of a fusion-fission hybrid energy reactor, in which ^{232}Th is mainly used to breed ^{233}U by capture reaction. It is essential to validate ^{232}Th nuclear data by carrying out integral fusion neutronics experiments for macroscopic thorium assemblies. The thorium assemblies with a D-T fusion neutron source consist of a polyethylene shell, depleted uranium shell, and thorium oxide cylinder. The activation of γ-ray off-line method for determining the thorium reaction rates is developed. The $^{232}Th(n, \gamma)$, $^{232}Th(n, f)$, and $^{232}Th(n, 2n)$ reaction rates in the assemblies are measured by using ThO_2 foils and an HPGe γ spectrometer. From ^{232}Th reaction rates, the fuel and neutron breeding properties of thorium under different neutron spectra are obtained and compared. The leakage neutron spectra from the ThO_2 cylinders are measured by a liquid scintillation detector. The experimental uncertainties are analyzed. The experiments are simulated by using the MC code with different evaluated data. The ratios of calculation to experimental values are analyzed.

Keywords: neutronics experiment, D-T fusion, thorium assembly, ^{232}Th reaction rate, neutron spectra, MC simulation

1. Introduction

The fusion-fission hybrid energy reactor, consisting of a low-power magnetic confinement fusion assembly and a subcritical blanket, is one of the advanced reactors of applying fusion technology to solve the present energy crisis. Natural thorium contains one isotope ^{232}Th. Thorium is a fertile element that can be applied in the conceptual blanket design of a fusion-fission hybrid reactor [1, 2]. The actual neutron spectrum in the subcritical blanket based on the Th/U fuel cycle is composed of fast and thermal spectra. The ^{232}Th capture cross section at fast neutron is slightly larger than that of ^{238}U, and ^{232}Th is more suitable to breed ^{233}U under fast spectrum. Since ^{232}Th capture cross section for thermal neutron is about 2.7 times larger than that of ^{238}U, the conversion rate in the Th/U fuel cycle is more than that in the U/Pu fuel cycle and the neutron economy of thorium is better. Moreover, the ^{233}U capture cross section for thermal neutron is smaller than that of ^{239}Pu and ^{233}U needs to absorb neutrons many times to produce Pu and long-life Minor Actinides (MA, such as ^{237}Np, ^{241}Am, and ^{242}Cm), whereas Pu and MA produced in the Th/U fuel cycle are one order of magnitude less than those in the U/Pu fuel cycle. Therefore, the Th/U fuel cycle is beneficial to reduce the long-life nuclear waste and prevent nuclear proliferation. The feasibility and reliability of the physical design for the subcritical blanket based on thorium depend on the accuracy of ^{232}Th nuclear data and calculational tool. It is essential to carry out the fusion neutronics experiments for validating the evaluated ^{232}Th nuclear data and studying the breeding properties.

A small number of fusion neutronics experiments on thorium were carried out, and there exist essential differences between the calculations and experiments [3–5]. The ^{232}Th fission rate with fast neutrons was determined by detecting the gamma rays emitted from ^{140}Ba and ^{140}La, and the calculated-to-experimental ratio was 0.9 based on ENDF/B-IV [4]. The thorium fission reaction rate in a metallic sphere setup was determined by absolute measurement of the gamma-emission from ^{143}Ce, the experimental uncertainty was 5.2%, and the calculation to experiment ratio was 1.17 employing ENDF/B-IV [5].

The integral fusion neutronics benchmark experiments for macroscopic thorium assemblies with a D-T fusion neutron source were carried out at Institute of Nuclear Physics and Chemistry (INPC) [6–17]. The method for measuring integral ^{232}Th reaction rate and its application in an experimental assembly were developed and investigated [6–8]. In this chapter, the progress in the fusion neutronics experiments for thorium assemblies is described. The overview of main results is presented. The thorium assemblies with a D-T fusion neutron source consist of a polyethylene shell, depleted uranium shell, and thorium oxide cylinder. The ^{232}Th reaction rates in the assemblies and leakage neutron spectra are measured separately. The benchmark experiments on fuel and neutron breeding properties derived from the ^{232}Th reaction rates in representative thorium assemblies are carried out and analyzed. The breeding properties are valuable to the breeding ratio in the conceptual design of subcritical blanket based on the Th/U fuel cycle. The experimental results are simulated by using the MC code with different evaluated data. The ratios of calculation to experimental values are analyzed.

2. Methods

The fusion neutronics experiments contain the measurements of the ^{232}Th(n,γ), ^{232}Th(n, f), and ^{232}Th(n,2n) reaction rates, and the neutron spectra for thorium assemblies with a D-T fusion neutron source.

2.1 ^{232}Th reaction rates

The experimental method of activation of γ-ray off-line measurement of ^{232}Th reaction rates is used. The activation γ-rays are measured by using an HPGe γ spectrometer.

The ^{232}Th capture reaction rate (THCR) indicates the fuel breeding, that is, the production rate of fissile ^{233}U (^{233}Pa decay). THCR can be deduced by measuring 311.98 keV γ rays emitted from ^{233}Pa [6, 7]. The reaction process is as follows:

$$^{232}Th \xrightarrow{(n,\gamma)} {}^{233}Th \xrightarrow{\beta^-,22.3\text{min}} {}^{233}Pa \xrightarrow{\beta^-,26.967d} {}^{233}U \quad (1)$$

The ^{232}Th fission (with threshold of 0.7 MeV) reaction rate (THFR) indicates energy amplification and neutron breeding. The fission fragment yield correction method is used [8]. THCR can be deduced by measuring 151.16 keV γ rays emitted from the decay of ^{85m}Kr from ^{232}Th (n, f) reaction. The reaction process is as follows:

$$^{232}Th \xrightarrow{(n,f),Y_f} {}^{85m}Kr \xrightarrow{\beta^-,4.48h} {}^{85}Rb \tag{2}$$

The ^{232}Th(n,2n) ^{231}Th (with threshold of 6.5 MeV) reaction rate (THNR) indicates neutron breeding. THNR is obtained from measuring 84.2 keV γ rays emitted from ^{231}Th [9]. The reaction process is as follows:

$$^{232}\mathrm{Th} \xrightarrow{(\mathrm{n,2n})} {}^{231}\mathrm{Th} \xrightarrow{\beta^-,25.52\mathrm{h}} {}^{231}\mathrm{Pa} \tag{3}$$

The ^{232}Th reaction rates are deduced from the measured activity and corrections, which include detection efficiency of the HPGe γ spectrometer, cited value of branching ratio, D-T neutron yield during irradiation, self-absorption of gamma rays in the foils, ^{85m}Kr yield only for THFR, etc. The ^{232}Th reaction rates are normalized to one source neutron and one ^{232}Th atom.

2.2 Breeding properties

The breeding ratio in the conceptual design of subcritical blanket is more than one [1]. The experiment on breeding properties of thorium is used to support the design [17]. The breeding properties are relevant to the reaction type, cross section, and neutron spectrum. The breeding properties contain the fuel breeding and neutron breeding. The fuel breeding is derived from the reaction rate ratio of ^{232}Th capture to fission, and neutron breeding from the ^{232}Th(n,2n) and fission reaction rates. The different neutron spectra are constructed by using the macroscopic assemblies in which the material is relevant to that of the conceptual design. The breeding properties under different assemblies are obtained and analyzed from the measured ^{232}Th reaction rates.

2.3 Neutron spectra

The neutron spectra leaking from the ThO_2 cylinders of different thickness are measured by the proton recoil method and the liquid scintillator [16]. The n-γ pulse shape discrimination is based on the cross-zero method. The spectra are resolved by using iterative method, and their range is from 0.5 to 16 MeV.

3. Assemblies

The experimental assemblies are composed of polyethylene shell, depleted uranium shell, and ThO_2 cylinder with a D-T fusion neutron source and thorium samples.

3.1 Polyethylene shell

One can assume the elastic scattering cross sections of H and C, which are widely used as standard cross sections [18] to be reliable. The polyethylene (PE) shell is adopted for checking the method of measuring the ^{232}Th reaction rates. The inner radius (IR) and the outer radius (OR) of the PE shell are 80 and 230 mm [11], respectively. Five slices of ThO_2 (concentration > 99.95%) foils are put in the radial channel at 0° to the incident D^+ beam, as shown in **Figure 1**. The mass and size of foils are about 4.2 g and ϕ30 × 1 mm, respectively.

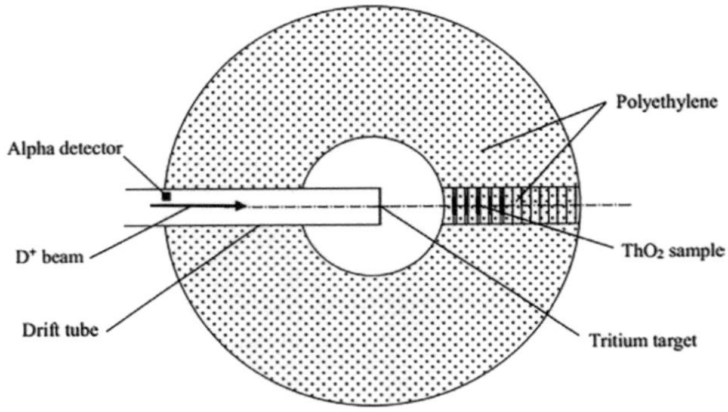

Figure 1.
Polyethylene shell assembly.

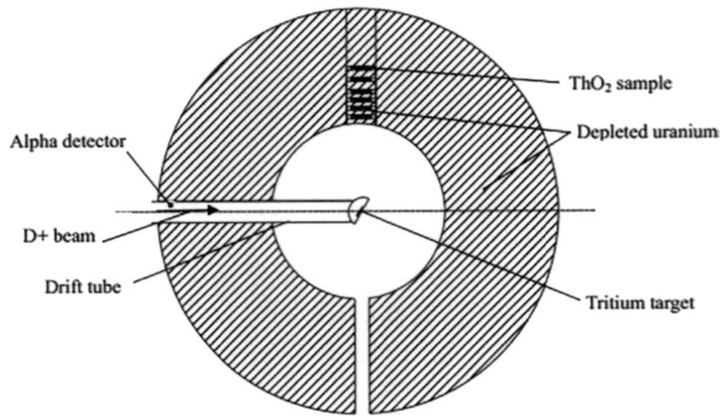

Figure 2.
Depleted uranium shell assembly.

A D-T fusion neutron source is located in the center of the shell. The 14 MeV neutrons are produced by a neutron generator at INPC. The energy of D^+ beam bombarding a T-Ti target is 225 keV. An Au-Si surface barrier semiconductor detector is at an angle of 178.2° to the incident D^+ beam in the drift tube and used to measure the absolute yield by counting associated α particles [19, 20]. D-T neutron yield is about 3×10^{10}/s.

3.2 Depleted uranium shell

In the conceptual design of a subcritical blanket based on thorium, the neutrons from the U reaction process are used to maintain the Th/U fuel cycle. The depleted uranium (DU) shell is adopted for studying Th reaction. The IR/OR of the DU shell is 131/300 mm [12]. Six slices of ThO_2 samples are put in the radial channel at 90° to the incident D^+ beam, as shown in **Figure 2**. ThO_2 samples are foils made from ThO_2 powder filling a plexiglass box with IR/OR of 9/9.5 mm. The mass of ThO_2 powder is about 0.45 g, and the thickness is about 0.7 mm. The D-T neutron source is located in the center of the shell.

3.3 ThO_2 cylinders

3.3.1 ThO_2/DU cylinders

The thorium oxide (ThO_2) cylindrical assembly with the thickness of 150 mm is produced and consists of three ThO_2 cylinders with the thickness of 50 mm and the diameter of 300 mm. The ThO_2 cylinders are made by pressing ThO_2 powder using PEO (CH_2CH_2O) as the binder and their densities are 4.25–5.59 g/cm^3 [9, 10]. The structure of the ThO_2 cylinders as benchmark is simple. To change neutron spectra in ThO_2 cylinders, the latter can be combined with DU cylinders. The combination of two ThO_2 cylinders and one DU cylinders is shown in **Figure 3**. Three slices of the ThO_2 samples are put in axial channel of the assembly. The front surface of the assembly is 113 mm from the center of a tritium target.

3.3.2 ThO_2 powder cylinder

Based on thorium oxide powder, the ThO_2 assembly is produced, as shown in **Figure 4** [13–15]. ThO_2 powder fills a stainless steel/aluminum cylinder container with IR/OR of 93.4/96.2 mm. The height of the ThO_2 cylinder is 168.9 mm and the density 1.5 g/cm^3. Five pieces of ThO_2 foils are put at 0° to the incident D^+ beam and fixed using holders consisting of aluminum plate and stainless steel. The mass and size of ThO_2 foils are about 5.0 g and ϕ30 × 1 mm, respectively. The distance between the tritium target center and the front end of the cylinder is 78.8 mm.

3.4 Neutron spectra in three assemblies

The neutron spectra in PE, DU, and ThO_2 assemblies are simulated by using the MCNP4B code [21] with ENDF/B-VII.0 [22], in which the S (α, β) thermal scattering model in PE is considered. The angular dependences of the source neutron

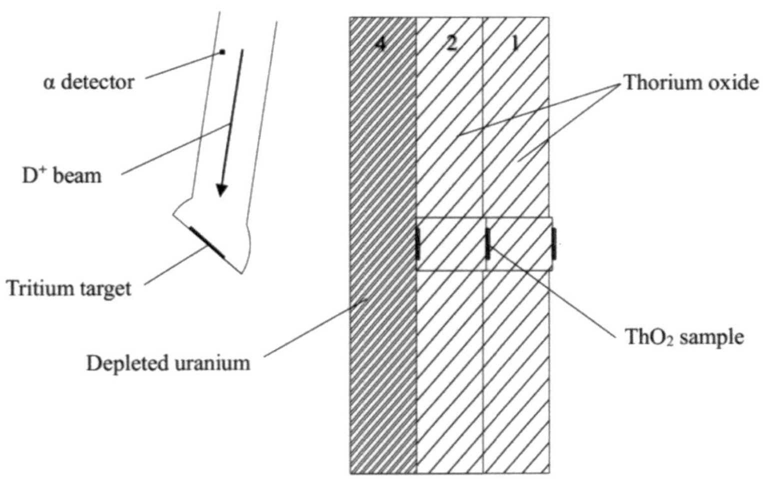

Figure 3.
ThO_2/DU assembly.

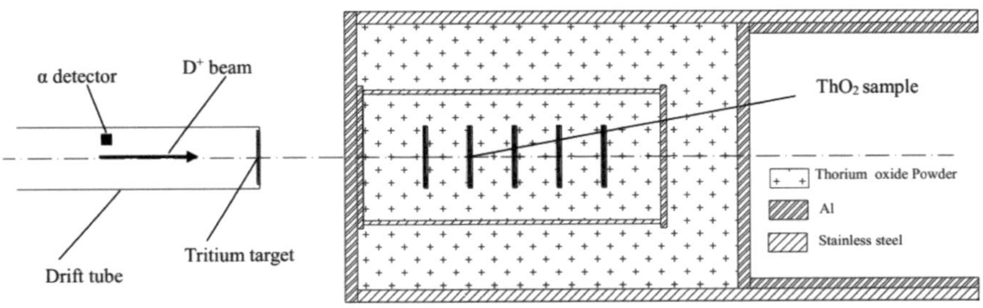

Figure 4.
ThO_2 powder cylindrical assembly.

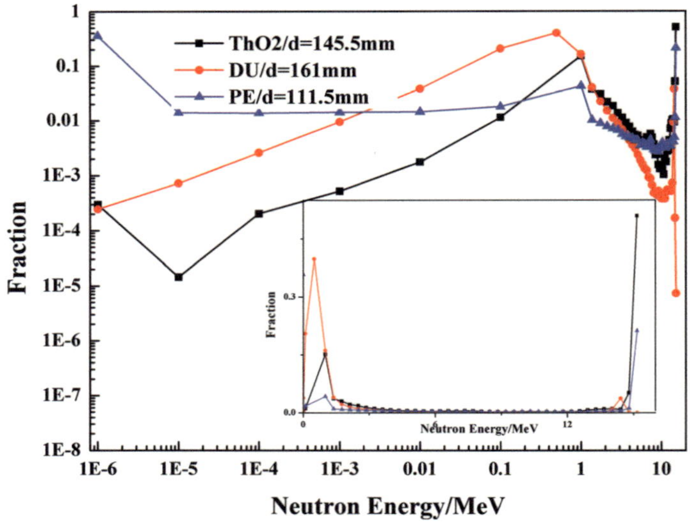

Figure 5.
Neutron spectra at foils in three assemblies.

energy and intensity are calculated by "DROSG-2000" code [23]. The neutron spectra at foils with different distances *d* to the neutron source in three assemblies are relatively compared, as shown in **Figure 5**. The ordinate is a normalized neutron fraction, that is, the proportion of the neutron number in each energy segment to the one in the whole energy range [11, 13]. The results show that the differences of the fractions are very obvious, especially in the low-energy region.

4. Results

4.1 ^{232}Th reaction rates in PE shell

The PE shell assembly for measuring ^{232}Th reaction rates is shown in **Figure 1**. THCR is deduced from measuring 311.98 keV γ rays emitted from ^{233}Pa (its half-life is 26.967 days, it is obtained from ^{233}Th decay). THFR is deduced from measuring 151.16 keV γ rays emitted from ^{85m}Kr decay (its half-life is 4.48 hour), which is one of the fragments of ^{232}Th(n,f) reaction, and using the fragment yield correction method. THNR is deduced from measuring 84.2 keV γ rays emitted from ^{231}Th (its half-life is 25.52 hour).

The experimental uncertainty of THCR is 3.1%, including neutron yield 2.5%, γ-ray detection efficiency 1.0% (HPGe-GEM 60P), self-absorption 1.0%, characteristic gamma branch ratio 1.0%, ^{232}Th nucleus number 0.5%, and counting statistics 0.3–0.6%.

The experimental uncertainty of THFR is 5.3%, including neutron yield 2.5%, γ-ray detection efficiency 1.0%, self-absorption 1.0%, average fission yield of ^{85m}Kr 4.3%, characteristic gamma branch ratio 0.7%, ^{232}Th nucleus number 0.5%, and counting statistics 0.8–1.0%.

The experimental uncertainty of THNR is 6.8%, including neutron yield 2.5%, γ-ray detection efficiency 1.0%, self-absorption 1.0%, characteristic gamma branch ratio 6.1%, ^{232}Th nucleus number 0.5%, and counting statistics 0.5–0.6%.

The experiment is simulated by using the MCNP code with evaluated nuclear data from different libraries, including ENDF/B-VII.0, ENDF/B-VII.1 [24] and JENDL-4.0 [25]. The model is completely consistent with the structure of the assembly; it takes into account the target chamber and experimental hall. The calculated statistical uncertainty is less than 1%. The ranges of C/E with ENDF/B-VII.0 are 0.96–1.02 for THCR, 0.95–0.97 for THFR, and 0.89–0.91 for THNR. The results show that the experiment and calculation for THCR and THFR are well consistent within the range of experimental uncertainties, respectively. It is shown that the γ-ray off-line method is feasible for determining the ^{232}Th reaction rates.

The distributions of ^{232}Th reaction rates obtained from the experiments and calculations with ENDF/B-VII.0 are shown in **Figure 6**. The reaction rate ratio of ^{232}Th capture to fission gives fissile production rate in unit of fuel burn-up [12]. The relative ratios measured are about 10.76–20.17 with the increase of radius in PE shell.

The ratios of calculation to experimental values (C/E) are analyzed. The C/E ratios of ^{232}Th reaction rates are shown in **Figure 7**, and the ^{232}Th(n,f) reaction results for different evaluated nuclear data are shown in Ref. [11]. The calculations with ENDF/B-VII.0 and ENDF/B-VII.1 for THNR underestimate the experimental values. Meanwhile, large differences still exist in the ^{232}Th(n,2n)^{231}Th cross sections among different evaluated data [26]. Fractions with different energies in the PE shell are calculated by using ENDF/B-VII.0, and neutrons of energy more than 6.5 MeV account for 33–48% in the whole energy range, as shown in **Figure 5**. Since the neutron spectra in the PE shell are reliable, it is suggested that ^{232}Th(n,2n) reaction cross sections should be studied further.

4.2 ^{232}Th reaction rates in DU shell

The DU shell assembly for measuring ^{232}Th reaction rates is shown in **Figure 2**. The ^{232}Th reaction rates are measured by the same method as described above.

The experimental uncertainties are 3.1% for THCR, 5.3–5.5% for THFR [6, 8], and 6.8% for THNR in DU shell.

The experiment is simulated using the MCNP code with different evaluated data, including ENDF/B-VII.0, ENDF/B-VII.1, JENDL-4.0, and CENDL-3.1 [27]. The distributions of ^{232}Th reaction rates from the experiments and calculations with ENDF/B-VII.0 are shown in **Figure 8**. The ranges of C/E ratios with ENDF/B-VII.0

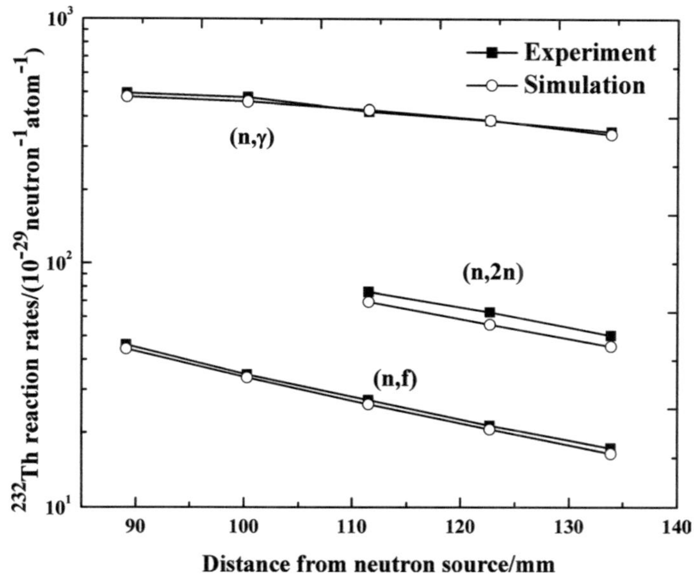

Figure 6.
^{232}Th reaction rates in PE shell.

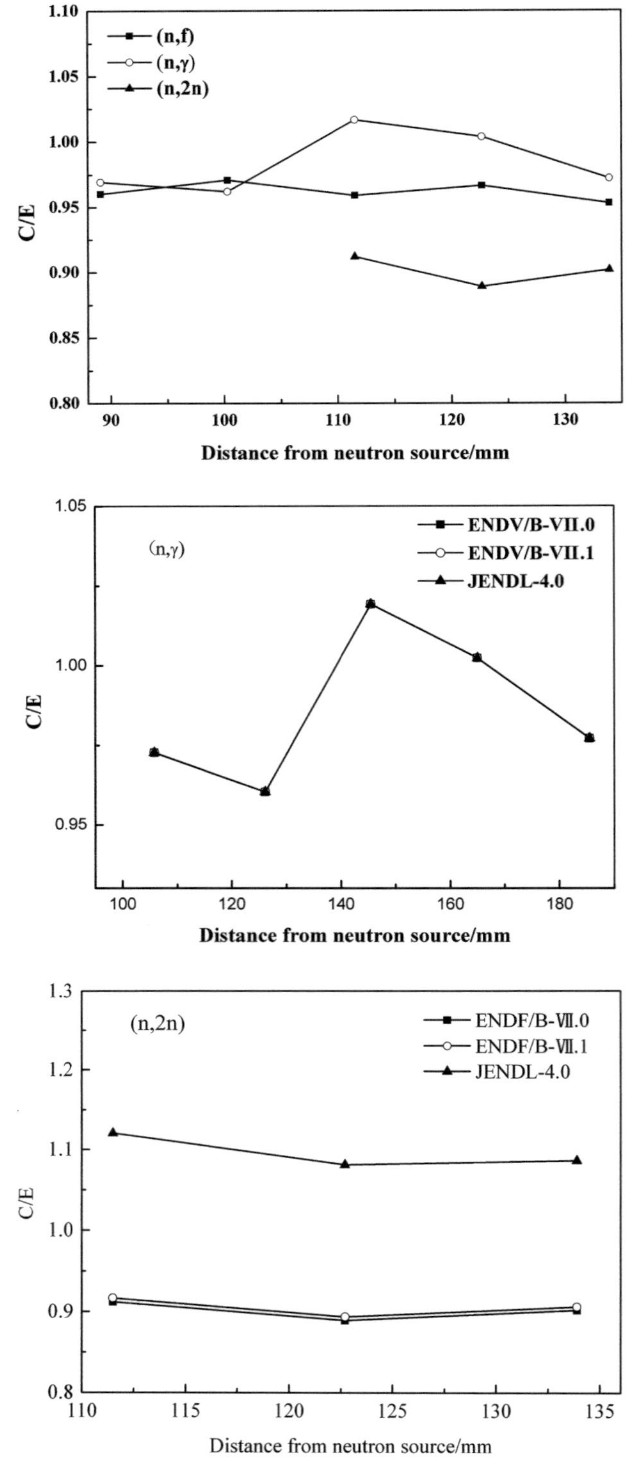

Figure 7.
C/E ratio of ^{232}Th reaction rates in PE shell.

are 0.97–1.04 for THCR and 0.95–1.02 for THFR [8, 12], respectively. The results show that calculations and experiments are well consistent within the range of experimental uncertainties. The ratio of ^{232}Th capture to fission is about 6.71–12.23 with the increase of radius in DU shell.

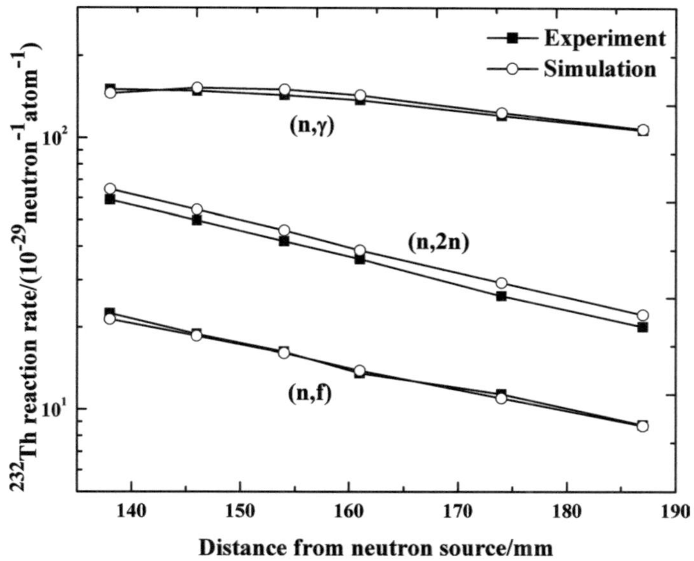

Figure 8.
^{232}Th reaction rates in DU shell.

The C/E ratios of ^{232}Th reaction rates with different evaluated data are shown in **Figure 9**. The calculations for THNR overestimate the experiments. Meanwhile, large differences still exist in C/E of THNR. The range of C/E with ENDF/B-VII.0 is 1.07–1.12. Fractions with different energies in DU shell are calculated by using ENDF/B-VII.0, and neutrons of energy more than 6.5 MeV account for 4–9% in the whole energy range, as shown in **Figure 5**. Since U(n,f) cross sections are standard in the wide energy range, it is suggested that U inelastic cross sections and ^{232}Th(n,2n) reaction cross sections should be studied further.

4.3 ^{232}Th reaction rates in ThO_2 cylinders

4.3.1 ^{232}Th fission and (n,2n) reaction rates in ThO_2 cylinder

The ThO_2 assembly for measuring ^{232}Th reaction rates in three ThO_2 cylinders with the thickness of 150 mm (without DU cylinder) is shown in **Figure 3**. The ^{232}Th fission and (n,2n) reaction rates are measured by the same method as described above.

The experimental uncertainties are 5.3–5.5% for THFR and 7.1% for THNR [9, 10].

The ^{232}Th reaction rates are calculated by using MCNP code with ENDF/B-VII.0. The ranges of C/E are 0.77–0.91 for THFR, and 0.92–1.0 [12] for THNR, respectively. The results show that the calculations generally underestimate the experiments for THFR. The PEO influence on THFR is described below. The distributions of ^{232}Th reaction rates by the experiments and calculations are shown in **Figure 10**.

4.3.2 ^{232}Th fission rates in ThO_2/DU cylinders

Experimental and simulative studies of THFR are carried out on three sets of ThO_2/DU cylinder assemblies to validate the evaluated thorium fission cross section and code [9, 10]. The size of each ThO_2 cylinder and DU cylinder is ϕ300 × 50 mm. The ThO_2 cylinders with PEO contents of 7.28, 1.1, and 0.55% are named as number 1, number 2, and number 3, respectively. The DU cylinder is named as number 4. Three sets of cylinder assemblies are combined with different cylinders, and named as "3 + 2 + 1," "4 + 2 + 1" (as shown in **Figure 3**) and "3 + 4 + 2 + 1" assembly, respectively.

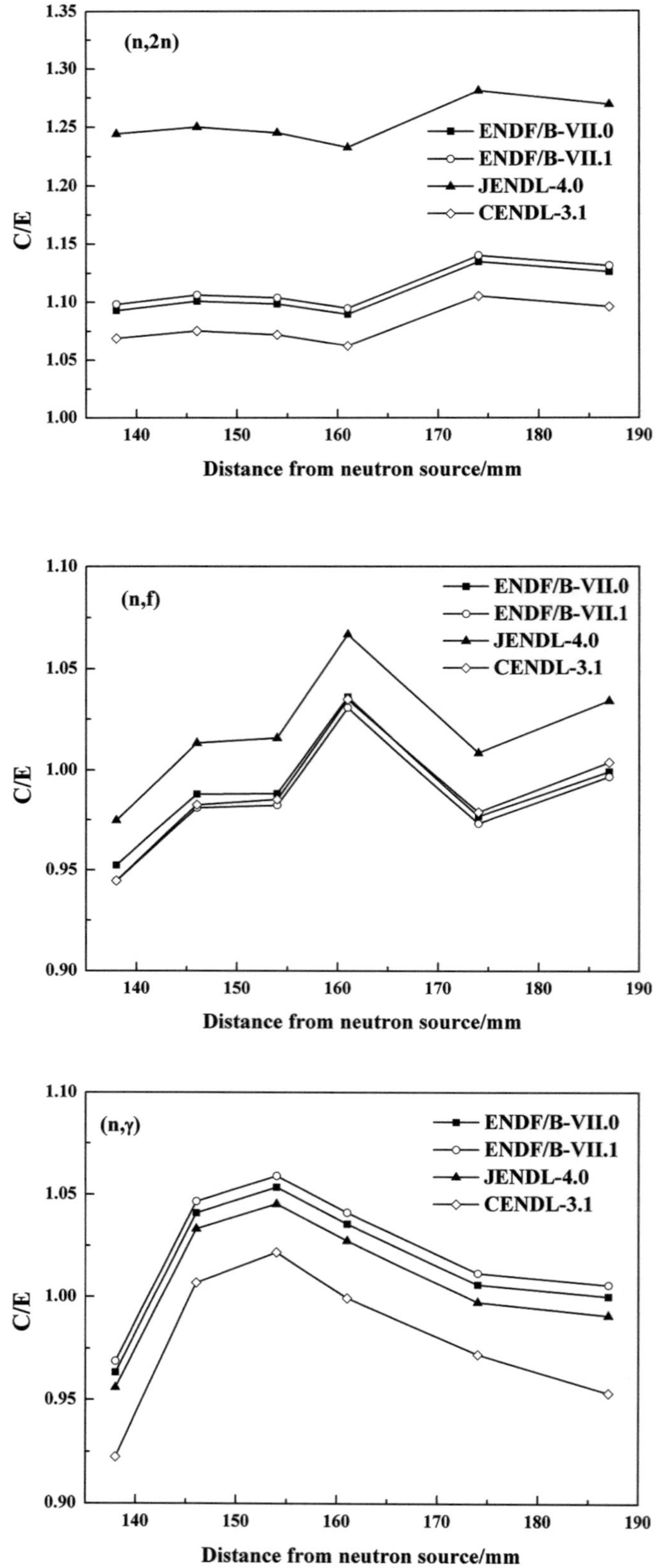

Figure 9.
C/E ratio of ^{232}Th reaction rates in the DU shell.

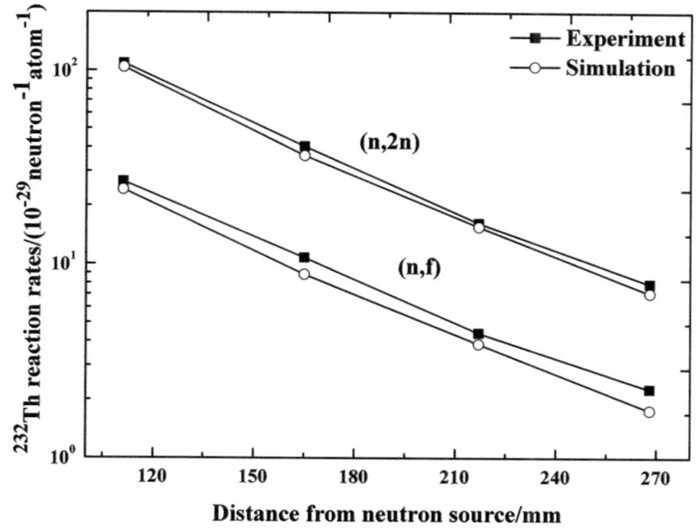

Figure 10.
^{232}Th reaction rates in ThO_2 cylinder.

THFR in the axial direction of the assemblies is obtained by using the activation method as described above, with experimental uncertainties about 5.6–5.9%.

THFRs are calculated by using MCNP code with ENDF/B-VII.0 and ENDF/B-VII.1. The calculations are 5–21% smaller than experimental ones, while the calculations with ENDF/B-VII.0 show better agreement with experimental ones. C/E distributions in the three assemblies are presented in **Figure 11**. The influence of the PEO in the ThO_2 cylinders is also evaluated by MCNP simulation employing ENDF/B-VII.0. The results show that the PEO influence on THFR under the measured level is negligible.

In order to gain more experimental results, it is necessary to design a new integral experiment employing thorium transport medium in which the ingredient is single and precisely known, and to determine THFR based on more kinds of fission

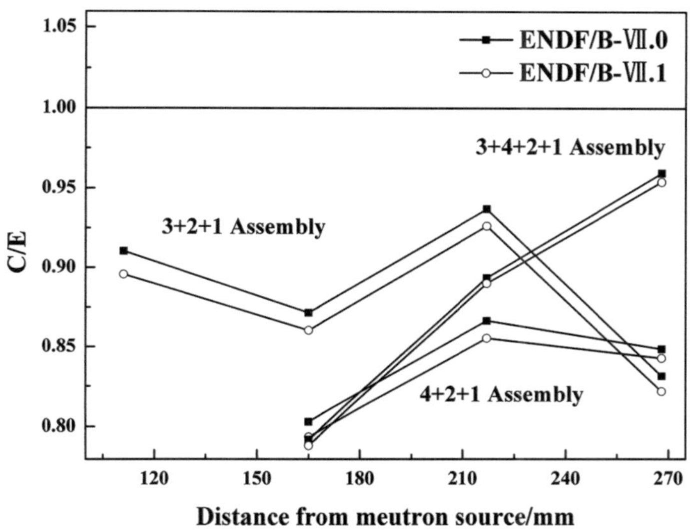

Figure 11.
C/E distribution in the three sets of assemblies.

products, as described below. The stage results could provide reference for the evaluation of neutron-induced thorium fission cross section, and the conceptual design margin of the subcritical blanket.

4.3.3 ^{232}Th reaction rates in ThO_2 powder cylinder

The ThO_2 power cylinder assembly for measuring ^{232}Th reaction rates is shown in **Figure 4**. The ^{232}Th reaction rates are measured by the same method as described above.

The experimental uncertainties are 3.1% for THCR, 5.5% for THFR, and 7.0% for THNR in the ThO_2 powder cylinder.

The experiment is simulated by using the MCNP code with different evaluated data [10, 11]. The C/E ratio of ^{232}Th reaction rates with ENDF/B-VII.0 are shown in **Figure 12**. The ranges of C/E ratio are 0.96–0.98 for THCR, 0.96–0.99 for THFR, and 0.74–0.76 for THNR. The results show that calculations and experiments for THCR and THFR are well consistent within the range of experimental uncertainties. The distributions of ^{232}Th reaction rates in the experiments and calculations are shown in [13–15].

The calculations for THNR underestimate the experiments. Fractions with different energies in ThO_2 powder cylinder are calculated by using ENDF/B-VII.0, and neutrons of energy more than 6.5 MeV account for 62–72% in the whole energy range, which is the largest among the assemblies, as shown in **Figure 5**. The suggestion described above is that ^{232}Th(n,2n) reaction cross sections should be studied further.

4.3.4 ^{232}Th fission rate based on ^{135}I in ThO_2 powder cylinder

The ThO_2 power cylinder assembly for developing the activation method of measuring THFR is shown in **Figure 4**. THFR in the axial direction of the cylinder is determined by measuring the 1260.409 keV gamma emitted from ^{232}Th fission product ^{135}I, with experimental uncertainties of 6.2% [14]. The experiment is simulated by using the MCNP code with ENDF/B-VII.0, ENDF/B-VII.1, JENDL-4.0, and CENDL-3.1. The calculations and experiments are in good agreement within experimental uncertainties. The activation method to determine THFR is developed

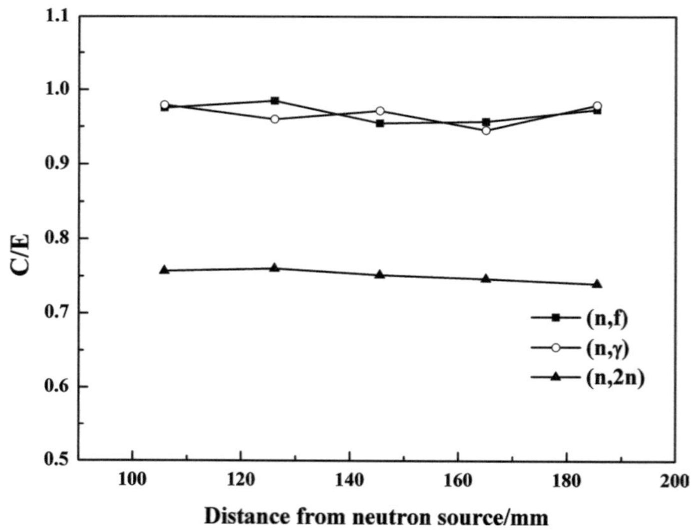

Figure 12.
C/E ratio of ^{232}Th reaction rates in ThO_2 powder cylinder.

and the data obtained in this work could provide reference for the validation of thorium fission parameters. The C/E ratio of ^{232}Th fission rates based on different evaluated data is presented in the [14].

4.4 Breeding properties

4.4.1 Fuel breeding

The primary conversion rate is one of the important parameters in the conceptual design of subcritical blanket. The relative reaction rate ratio of ^{232}Th capture to fission as the fissile production rate indicates fuel breeding in the fuel burn-up unit [12]. The ratios of ^{232}Th capture to fission measured in PE shell, DU shell, and ThO_2 powder cylinder are obtained.

The ratios are about 10.76–20.17 with the increase in radius of the PE shell. It is demonstrated that the fuel breeding efficiency under the neutron spectra in the PE shell is quite high.

The ratios are about 6.71–12.23 with the increase in radius of the DU shell. It is demonstrated that the fuel breeding efficiency under the neutron spectra in DU shell is high.

The ratios are only about 0.11–0.19 with the increase in radius of the ThO_2 powder cylinder. It is demonstrated that the fuel breeding efficiency under the neutron spectra in ThO_2 powder cylinder is low.

The results show that the ratios are relevant to neutron spectra in the assemblies. The ratios in the three assemblies are compared and shown in **Figure 13**.

4.4.2 Neutron breeding

The bred neutrons from ^{232}Th(n,2n) and ^{232}Th(n,f) react with thorium or relevant nuclides to maintain the Th/U fuel cycle. THNRs in three assemblies, that is, under different neutron spectra, are compared and shown in **Figure 14**. The results show that the ^{232}Th(n,2n) reaction rates are relevant to the fraction of high-energy neutrons in the assemblies as described above, and the decreasing trend of THNR with the increase in distance to the neutron source are similar for three assemblies.

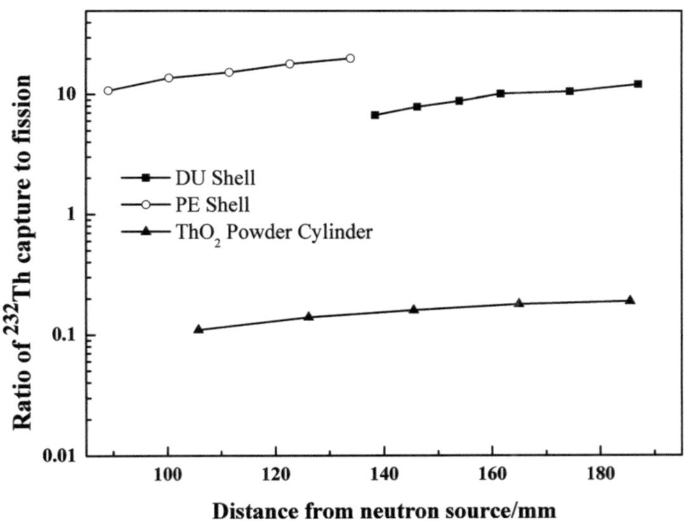

Figure 13.
Ratios of ^{232}Th capture to fission in the three assemblies.

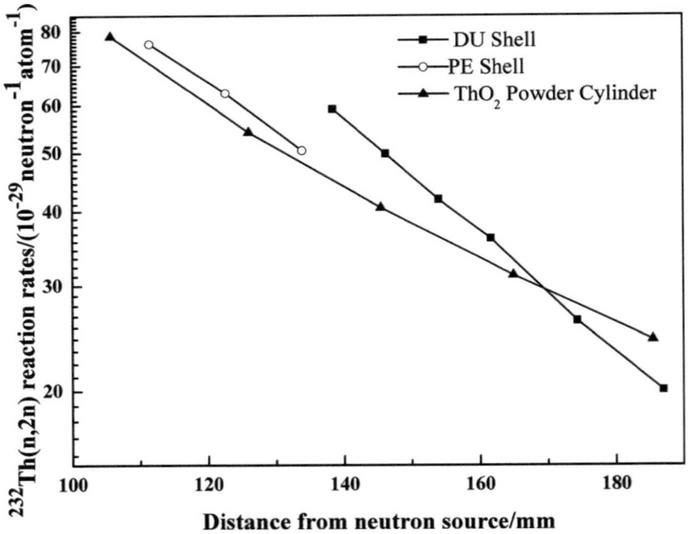

Figure 14.
THNRs in the three assemblies.

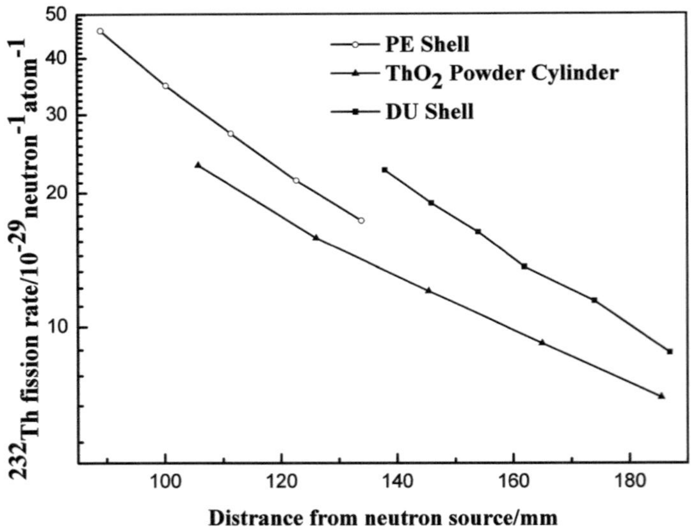

Figure 15.
THFRs in the three assemblies.

Since ^{230}Th half-life (7.54×10^4 years) is very long, measurement of $^{232}Th(n,3n)$ ^{230}Th (with threshold of 11.6 MeV) reaction rate by the activation method is very difficult. The $^{232}Th(n,4n)$ reaction has high threshold 19 MeV and is not involved in this work.

The prompt neutron and delayed neutron yields from $^{232}Th(n,f)$ reaction are about 3.7 and 0.0265 per fission at 14.1 MeV [28], respectively. THFRs in three assemblies, that is, under different neutron spectra, are compared and shown in **Figure 15**. From **Figures 14** and **15**, THNRs are higher than THFRs in the three assemblies.

4.5 Leakage neutron spectra

Three assemblies consist of the ThO_2 cylinders with thicknesses of 50, 100, and 150 mm (without DU cylinder), respectively, as shown in **Figure 3**. The front

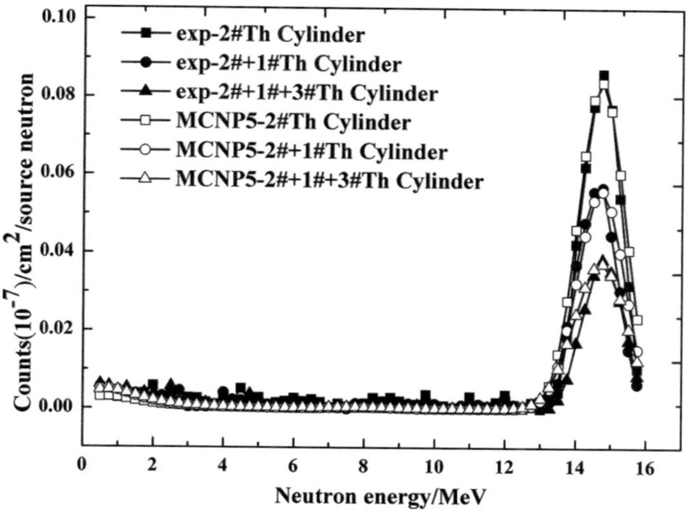

Figure 16.
Leakage neutron spectra from ThO_2 cylinders.

surface of the assembly is 0.22 m from the center of a T-Ti target. The leakage neutron spectra are measured by using a 50.8 mm diameter and 50.8 mm length BC501A liquid scintillator coupled to a 50.8 mm diameter 9807B photomultiplier [16]. The distance from the detector to the neutron source is 10.75 m. The detector is at a 0° to the incident D^+ beam and arranged in shielding room. The influence of background neutrons is negligible.

The leakage neutron spectra from the three assemblies are measured. The spectra are normalized to one source neutron and unit area. The experimental uncertainties are 9.7% for 0.5–1 MeV, 6.7% for 1–3 MeV, and 6.3% for 3–16 MeV. The experiments are calculated by using MCNP code with ENDF/B-VII.0. The results show that the experiments and calculations are generally consistent within the range of experimental uncertainties, and the spectra (<5 MeV) should be analyzed further, as shown in **Figure 16**.

5. Conclusions

To validate ^{232}Th nuclear data, the fusion neutronics experiments for the three kinds of thorium assemblies with a D-T neutron source have been carried out. The two spherical assemblies based on the DU and PE shells, and the cylindrical assemblies based on ThO_2 have been designed and established. The assembly materials are referable to the conceptual design of subcritical blanket of a hybrid reactor. The ^{232}Th(n,γ), ^{232}Th(n,f), and ^{232}Th(n,2n) reaction rates in the assemblies are measured by the foil activation technique. The results show that the developed activation approach can work well for the experiments, and the ^{232}Th reaction rates are relevant to neutron spectra in assemblies. The reaction rate ratios of ^{232}Th capture to fission are obtained. The fuel and neutron breeding properties under different neutron spectra are compared and analyzed. The leakage neutron spectra from ThO_2 cylinders are measured. The experimental results are compared to the numerical results calculated by using the MCNP code with different evaluated data. The results show that the experiments are benefit to validate Th nuclear data and support the conceptual design of subcritical blanket with thorium in a hybrid reactor. Furthermore, it should be beneficial to measure relevant ^{232}Th excitation curve at white neutron source of China Spallation Neutron Source (CSNS) [29] for verifying ^{232}Th nuclear data.

Acknowledgements

This work is supported by the National Special Magnetic Confinement Fusion Energy Research of China (No. 2015GB108001B), the National Natural Science Foundation of China (No. 11675155, 91226104), and the National Key Research and Development Program of China (No. 2016YFA0401603). The author wishes to acknowledge all participators of the projects, including Dr. Yiwei Yang, Dr. Lei Zheng, Dr. Song Feng, MS. Caifeng Lai, Prof. Xinxin Lu, MS. Zhujun Liu, Prof. Li Jiang, Prof. Mei Wang, MS. Zijie Han, et al. All participators would like to thank Prof. Benchao Lou and his group for operating the neutron generator. The author thanks the reviewers, comments and suggestion.

Author details

Rong Liu
Institute of Nuclear Physics and Chemistry, Key Laboratory of Neutron Physics, China Academy of Engineering Physics, Mianyang, China

*Address all correspondence to: liurongzy@163.com

IntechOpen

© 2018 The Author(s). Licensee IntechOpen. This chapter is distributed under the terms of the Creative Commons Attribution License (http://creativecommons.org/licenses/by/3.0), which permits unrestricted use, distribution, and reproduction in any medium, provided the original work is properly cited.

References

[1] Shi X, Peng XJ. Preliminary concept design on blanket neutronics of a fusion-fission hybrid reactor for energy production. Nuclear Power Engineering. 2010;**31**(4):5-7. In Chinese

[2] Zhao J, Yang YW, Zhou ZW, et al. Study of thorium-uranium based molten salt blanket in a fusion-fission hybrid reactor. Fusion Engineering and Design. 2012;**87**(7):1385-1389

[3] Adam J, Bhatia C, Katovsky K, et al. A study of reaction rates of (n, f), (n, γ) and (n, 2n) reactions in ^{nat}U and ^{232}Th by the neutron fluence produced in the graphite set-up (GAMMA-3) irradiated by 2.33 GeV deuteron beam. European Physical Journal A: Hadrons and Nuclei. 2011;**47**(7):1-18

[4] Anderl RA, Harker YD. Measurement of the integral capture and fission cross sections for ^{232}Th in the CFRMF. In: Proceedings of the International Conference on Nuclear Cross Sections for Technology. Vol. 594. Tennessee, USA; 1980. p. 475

[5] Zagryadskii VA, Markovskii DV, Novikov VM, et al. Calculated neutron transport verifications by integral 14 MeV-neutron source experiments with multiplying assemblies. Fusion Engineering and Design. 1989;**9**(3):347-352

[6] Yang YW, Liu R, Yan XS, et al. Thorium capture ratio determination through γ-ray off-line method. Acta Physica Sinica. 2013;**62**(3):032801. In Chinese

[7] Yang YW, Liu R, Jiang L, et al. Determination of ^{232}Th(n, γ) reaction rate induced by D-T neutrons in one-dimensional alternate depleteduranium/polyethylene shells. Acta Physica Sinica. 2014;**63**(16):162801. In Chinese

[8] Feng S, Liu R, Lu XX, et al. Determination of thorium fission rate by off-line method. Acta Physica Sinica. 2014;**63**(16):162501. In Chinese

[9] Feng S, Yang YW, Lu XX, et al. An integral experiment on thorium oxide/depleted uranium cylinders with D-T neutrons for ^{232}Th(n, 2n) reaction. Annals of Nuclear Energy. 2015;**81**:281-286

[10] Zheng L, Lu XX, Yang YW, Liu R, et al. Experimental and simulative studies of thorium fission rateson thorium oxide/depleted uranium cylinders with D-T neutrons. Progress in Nuclear Energy. 2017;**99**:73-80

[11] Zheng L, Yang YW, Liu ZJ, et al. Measurement and analysis of thorium fission rate in a polyethylene shell with a D-T neutron source. Fusion Engineering and Design. 2016;**113**:177-182

[12] Liu R, Yang YW, Yan XS, et al. Measurement and calculation of U and Th reaction rates in uranium mock assemblies. Annals of Nuclear Energy. 2016;**92**(2):391-396

[13] Zheng L, Liu ZJ, Yang YW, et al. Measurement of Th(n,f) and Th(n,γ) reaction rates in thorium powder cylinder bombarded with D-T neutrons. Journal of Nuclear Science and Technology. 2017;**54**(5):600-608

[14] Zheng L, Yang YW, Liu R, et al. Determination of thorium fission rate based on ^{135}I in thorium oxide cylinder bombarded with D-T fusion neutrons. Annals of Nuclear Energy. 2018;**119**:264-270

[15] Liu ZJ, Yang CW, Yang YW, et al. Measurement and analysis of ^{232}Th(n,2n) reaction rate in the thorium oxide cylinder with a D-T neutron source. Annals of Nuclear Energy. 2018;**111**:660-665

[16] Liu R, Yang YW, Zheng L, et al. Integral experiments on thorium assemblies with D-T neutron source. In: EPJ Web of Conferences (ND2016). Vol. 146. 2017. p. 06022. DOI: 10.1051/epjconf/201714606022

[17] Liu R, Yang YW, Zheng L, et al. Benchmark experiments on breeding properties of thorium. Fusion Engineering and Design. 2018;**131**:119-124

[18] Carlson AD, Pronyaev VG, Smith DL, et al. International evaluation of neutron cross section standards. Nuclear Data Sheets. 2009;**110**(12):3215-3324

[19] Liu R, Lin LB, Wang DL, et al. Measurement and check of fusion neutron yield with the method of associated particles at a large angle. Nuclear Electronics and Detection Technology. 1999;**19**(6):428-432. In Chinese

[20] Yan J, Liu R, Li C, et al. LabVIEW-based auto-timing counts virtual instrument system with ORTEC 974 counter/timer. Journal of Nuclear Science and Technology. 2009;**20**(5):307-311

[21] Briesmeister JF. MCNP: A General Monte Carlo N-Particle Transport Code. LA-12625-M Version 4B (Issued)1997. p. 1

[22] Chadwick MB, Oblozinsky P, Herman M, et al. ENDF/B-VII.0: Next generation evaluated nuclear data library for nuclear science and technology. Nuclear Data Sheets. 2006;**107**(12):2931-3060

[23] Drosg M. DROSG-2000: Neutron Source Reactions. IAEA; 2003

[24] Chadwick MB, Herman M, Oblozinsky P, et al. ENDF/B-VII.1 nuclear data for science and technology: Cross sections, covariances, fission product yields and decay data. Nuclear Data Sheets. 2011;**112**(12):2887-2996

[25] Shibata K, Iwamoto O, Nakagawa T, et al. JENDL-4.0: A new library for nuclear science and engineering. Journal of Nuclear Science and Technology. 2012;**48**(1):1-30

[26] Reyhancan IA. Measurements and model calculations of activation cross sections for $^{232}Th(n,2n)^{231}Th$ reaction between 13.57 and 14. 83 MeV neutrons. Annals of Nuclear Energy. 2011;**38**:2359-2362

[27] Ge ZG, Zhao ZX, Xia HH, et al. The updated version of Chinese evaluated nuclear data library (CENDL-3.1). Journal of the Korean Physical Society. 2011;**59**(2):1052-1056

[28] Meadows J et al. Evaluated Nuclear Data File of Th-232. ANL/NDM-35. 1978

[29] The CSNS. Back-n collaboration: Back-n white neutron facility for nuclear data measurements at CSNS. Journal of Instrumentation. 2017;**12**(7):P07022

Section 2

Selected Problems of Mechanical Engineering

Chapter 4

Mechanical Mockup of IFE Reactor Intended for the Development of Cryogenic Target Mass Production and Target Rep-Rate Delivery into the Reaction Chamber

Irina Aleksandrova, Elena Koresheva, Evgeniy Koshelev, Boris Kuteev and Andrei Nikitenko

Abstract

Target production and its delivery into the reaction chamber of high repetition rate facilities are the most challenging issues in inertial fusion energy (IFE) research. At the Lebedev Physical Institute of Russian Academy of Sciences (LPI), efforts are underway on creation of the mechanical mockup of IFE reactor (MM-IFE) for developing the reactor-scale technologies applicable to mass production of IFE targets and their delivery with a repeatable rate into the chamber of IFE reactor. In this chapter, we discuss the current status and further trends of developments in the area of advanced target technologies underlying the research and development program on MM-IFE.

Keywords: inertial fusion energy, free-standing target systems, mass production, repeatable operation, noncontact delivery

1. Introduction

The goal of IFE research is development of high-precision and mass production technologies for fueling a commercial power plant at the rate of ~10 Hz [1]. The conventional approach to solid layering based on the beta-layering method [2] is unable to ensure the IFE requirements, as it (a) works with targets fixed on a suspension (no repetition rate operation), (b) has a long layering time (more than 24 h that leads to a large tritium inventory), (c) shows the grain boundaries dynamic under thermal and mechanical loads in time between the moment just after target preparation and the laser shot, which results in roughening of the layer surface and may lead to implosion instabilities, and (d) has a high production cost (more than $1000/target).

The beta-layering method can form a spherical fuel layer in a uniform thermal environment; however, it is inefficient in preventing local defects. Therefore, the

IntechOpen

modern requirements are asking for development of structure-sensitive methods aimed at new layering techniques meeting the IFE needs. This is due to the fact that the progress in plasma implosion up to intensive fusion reactions lies in formation of a given fuel structure that must be isotropic for reaching fusion conditions.

At the LPI, the concept of a mechanical mockup of IFE reactor has been proposed [3] to develop reactor-scale technologies applicable to mass production of IFE targets at significant rates (**Figure 1**). The LPI program also includes extensive development work on creation of different designs of the hybrid accelerators for IFE target transport with levitation (noncontact acceleration systems).

The MM-IFE is a modular facility representing in essence a free-standing target (FST) transmission line (integral part of any fusion reactor) designed to produce IFE targets and to provide their noncontact delivery at the laser focus and synchronous irradiation by a laser (1–10 Hz). It consists of 3 main blocks: (1) cryogenic target factory (CTF) operating with isotropic fuel layers of 200–300 μm thick (**Figure 1a**) [1]; (2) cryogenic IFE-target injector operating at accelerations <500 g and injection velocities $V_{inj} \geq 200$ m/s [4–8]; and (3) tracking systems for online characterization and control of the injected targets [9–12]. Replacement of the FST-layering module, being the main part of CTF, on the extruder of the solid fuel pellets allows developing the technologies for continuous fuel supply into magnetic fusion energy (MFE) facilities (**Figure 1b**).

Basic elements of the MM-IFE have been tested at LPI as prototypes for risk minimization at the stages of MM-IFE construction and startup. We especially highlight that moving targets are the necessary condition for realizing the repeatable target production at required rates, their mass manufacturing and noncontact delivery.

In this chapter, we discuss some challenging scientific and technological issues associated with IFE targets, the current status of the R&D program on MM-IFE, and further trends in developing the advanced IFE technologies for high-repetition rate laser facilities.

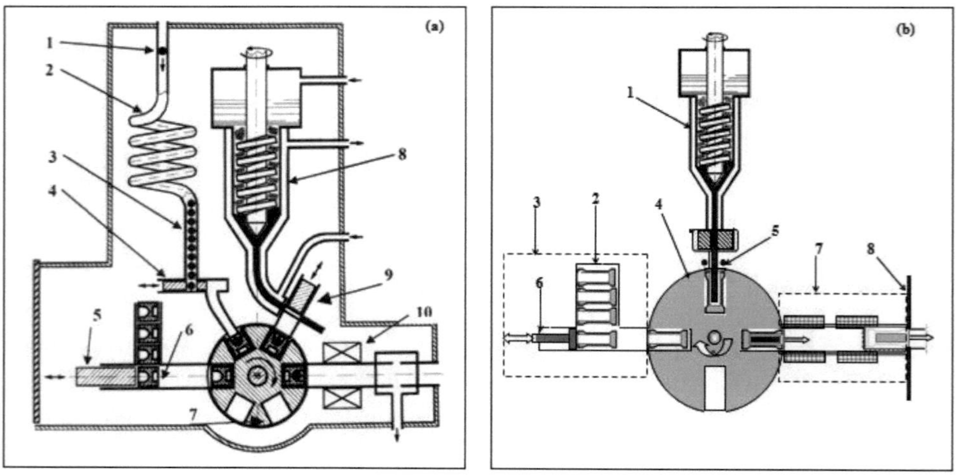

Figure 1.
Repeatable and mass production of the fuel targets/pellets for IFE (a) and for MFE (b) reactors. In (a): 1—fuel-filled polymer shell, 2—FST-layering module, 3—cryogenic targets batch, 4—shuttle, 5—pusher, 6—target carrier (sabot), 7—drum of a revolver type, 8—extruder for protective cover production, 9—coin for the protective cover formation and delivery, 10—field coil for "sabot + target + protective cover" pull out and delivery to the start point of injector. In (b): 1—extruder of solid fuel pellets, 2—pellet carrier (sabot), made from superconductor or ferromagnetic, 3—module for sabot-repeatable delivery to the rotating drum, 4—rotating drum for rep-rate assembly of the units "sabot + pellet," 5—heater for pellets production, 6—pusher, 7—linear electromagnetic accelerator (injector), 8—sabot brake.

2. Target mass production

The fuel structure is very important for the progress toward ignition. Considering the issue of high-quality cryogenic layer fabrication, we have to rely, first of all, on structural properties of hydrogen isotopes and their mixtures. Survivability of the fuel layers subjected to the environmental effects during target delivery may depend on the layer structure as well.

2.1 Structure-sensitive methods

Many important properties of materials are structure-sensitive, and often a relatively small number of defects have a disproportionately large effect on the material properties. Material structuring is very promising for creation of fusion materials with required properties. The role of structure-sensitive methods when developing new functional materials is especially underlined in [13]: "All materials have different chemical composition, aggregate states (solid body, liquid, and gas), allotropic modifications (graphite-diamond), or can be a mix of several substances (clay) that can be written as definition (1):

$$\begin{aligned}&\text{MATERIAL} = \\ &[\text{Chemical composition (A)} + \text{Aggregate state (A)} + \text{Allotropic modification (A)}] + \\ &[\text{Chemical composition (B)} + \text{Aggregate state (B)} + \text{Allotropic modification (B)}] + \ldots\text{''}\end{aligned} \tag{1}$$

However, this is obviously an inexact characteristic. The natural or introduced imperfection in the material is more important. Defects or, more generally, microstructure, define many major structure-sensitive properties of the materials. Critical for material property parameters are the type of available defects, their spatially organized packing, and interaction at multiple hierarchy levels. Chemical composition, aggregate state, and allotropic modification as a cooperative characteristic can be replaced with a more common concept "Phase." Then, the previous definition (1) can be transformed in (2):

$$\text{Material} = \text{Phase(s)} + \text{Microstructure} \tag{2}$$

The defect structure plays an important role in determining many material properties. From technological applications, a particular interest is the mechanical and thermal characteristics. In the IFE, a practical tool for correlating the structure and properties of the hydrogen fuel is the thermal target environment; cooling rates, fuel doping, periodic mechanical influence on the target under the cryogenic layer freezing, etc. [1]. For example, depending on the cooling rates, the solid fuel layer can be in the state with a different microstructural length or grain size: isotropic ultrafine layers and anisotropic molecular crystals (real single crystals, coarse-grained crystalline). This becomes particularly important if one takes into account the properties of the hydrogen isotopes.

In the equilibrium state, the solid hydrogen isotopes consist of anisotropic molecular crystals. In our analysis, we are guided by the fact that the angular dependence of the sound velocity, V, is known for a number of substances crystallizing in the hexagonal close-packed (hcp) phase. As found in [14], the sound velocity anisotropy is inherent to hcp-H_2 and hcp-D_2 as well, and makes nearly 20% (longitudinal sound) and 33% (transverse sound). In accordance with the Debye theory, the lattice thermal conductivity is directly proportional to the value of V.

Then, even in the case of a uniform thermal environment on the target surface of the anisotropic single-crystal layers, there is a difference in the radial target temperature in time. Therefore, under uniform target heating or cooling, the normal temperature gradients onto the inner surface of such layers become different. This initiates the spherically asymmetrical sublimation (or condensation) of fuel in the target cavity and results in the layer degradation with respect to roughness and thickness.

A conventional approach to solid layering (known as beta-layering method [2]) involves crystallization from a single-seed crystal in the fixed target under extremely slow cooling ($q \sim 3 \times 10^{-5}$ K/s) and precise cryogenic temperature control (<100 μK) for obtaining the layers like a single crystal. In a uniform thermal environment, the beta-layering method can form a spherical fuel layer, but it is not efficient in preventing the local defects. The target lifetime (layer roughness is less than 1-μm rms) is of a few seconds after reaching the desired temperature [2]. This is implication of the fact that D-T layers formed by beta-layering are obtained as a result of almost equilibrium process of the crystal growth, and all the features of the equilibrium crystalline state will be inherent in such layers, including the temperature-dependent behavior of the local defects on the inner D-T surface. Besides, the total layering time is ~24 h or even more [2]. Thus, the beta-layering method is not efficient for mass target fabrication for IFE. It is of one-of-a-kind technique, and this is very expensive [2].

In [1], it is shown that the fuel structure dominates among the important remaining risk factors because the progress in plasma implosion up to intensive fusion reactions lies in formation of the fuel structures which must be isotropic for reaching the fusion conditions. **Figure 2** illustrates the cooling rates required to obtain isotropic fuel layers to withstand the thermal and mechanical environment during target fabrication and delivery.

2.2 FST layering method for high-repetition rate facilities

The LPI program on the target science and technology has recently been focused on the ability to inexpensively fabricate large quantities of targets by developing a specialized layering module of repeatable operation. The targets must be free-standing, or unmounted. At the LPI, the experience gained in the technology development based on rapid fuel layering inside moving free-standing targets (which refers to as FST layering method) can be used for creation of a next-generation FST layering module for high-repetition rate facilities. For typical shell sizes (1–4 mm), the FST-layering time is very small (<30 s, see Section 2.2) in comparison with the beta-layering method usually applied in the compression experiments (≥24 h [2]).

FST layering is a structure-sensitive method to govern the fuel layer microstructure. Such approach has been developed at LPI [1] to form an isotropic ultrafine solid fuel (submicron crystalline called "fine-grained" crystalline and nanocrystalline) which supports the fuel layer survivability under target injection

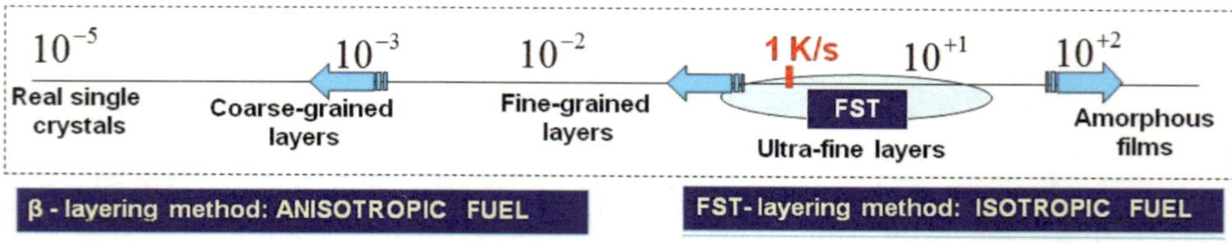

Figure 2.
The fuel structure of the cryogenic layers depends on the cooling rates, and so on the layering method.

and transport through the reaction chamber. During the FST layering, two processes are mostly responsible for maintaining a uniform layer formation:

- first, during the target rolling along the spiral layering channel, the forced target rotation results in a liquid layer symmetrization;
- second, the heat transport outside the target via conduction through a small contact area between the shell wall and the wall of the layering channel (metal hollow tube helium cooled outside) results in a liquid layer freezing.

Figures 3 and **4** show the operational scenario of the FST layering module (FST-LM):

- it works with a target batch at one time at cooling rates of q = 1–50 K/s;
- FST layering does not require the target surface to be near to isothermal;
- targets remain unmounted (or free-standing) in each production step;
- transport process is the target injection between the basic units: shell container (SC), layering channel (LC), test chamber (TC);
- SC holds the fuel-filled shells until the beginning of the FST layering process;

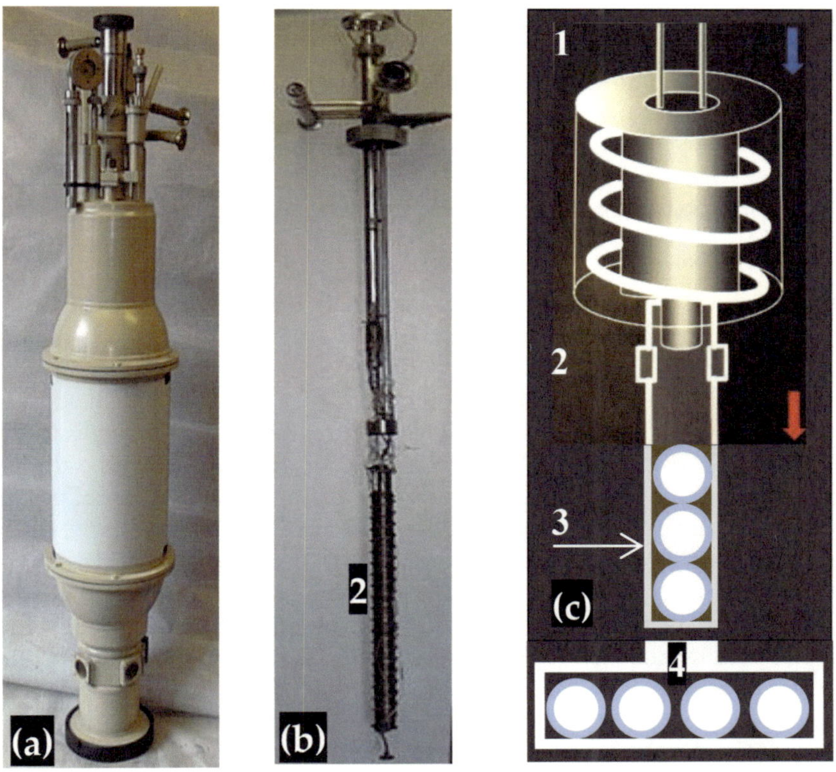

Figure 3.
FST-LM: (a) overview, (b) single spiral LC in assembly, (c) schematic of the FST-LM maintaining repeatable operation: 1—gravitation loading of the target with liquid fuel from SC to LC, 2—LC, 3—vertical target collector, 4—horizontal target collector.

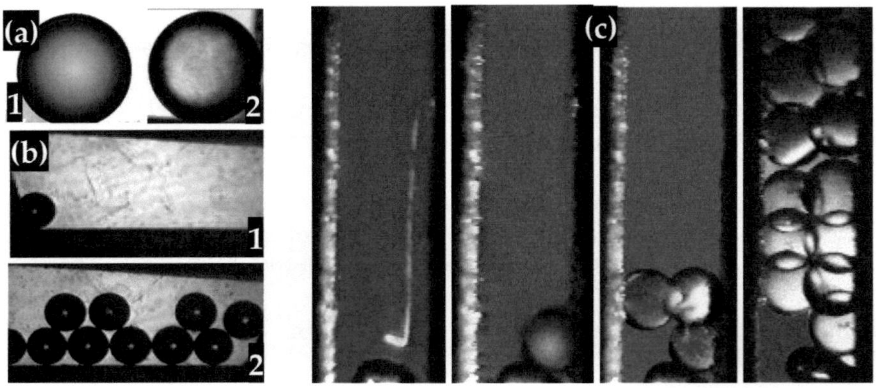

Figure 4.
The FST-LM operation with a rate of 0.1 Hz at T = 5.0 K. In (a) FST-layering results: 1—40-μm-thick layer (D_2 + 20% Ne) in the CH shell of 1.23-mm diameter (20% Ne-additives are used for modeling T_2 in the D-T fuel), 2—44-μm-thick H_2 layer with (H_2 + 5% HD) in the CH shell of 1.2-mm diameter; in (b) horizontal TC: 1—1 target (0 s), 2—10 targets (100 s); in (c) vertical TC: 12 targets in 100 s.

- LC is a special insert into the layering module cryostat; it is manufactured as a spiral (cylindrical or conical);
- combined LC (CLC) is one of the most interesting cases, which consists of two spirals (acceleration spiral and deceleration spiral) in order to reduce practically to zero the target speed at the CLC output;
- TC has two types of target collectors: vertical and horizontal;
- targets move top-down in the LC in a rapid succession of one after another that allows to realize a high injection rate during finished target delivery to the TC (**Figure 3**);
- TC is a prototypical interface unit between the layering module and target injector;
- TC is currently used for finished target quality control using precise tomographic and threshold characterization.

The goal of the target characterization program is to provide reliable data in an available time. In this respect, two technologies are important: (1) fuel layering technique development (detailed information about the spatial modes, which break the target symmetry) and (2) reaction chamber fueling of a commercial power plant (targets must be injected at significant rates (~10 Hz) which indicates evidence of a threshold behavior of the characterization process).

The first technology requires enhancing the obtained data, and the second needs shortening the characterization time. This indicates that the process details for the characterization technology dealing with the operational times and information content are of critical importance.

A hard development work is needed to take into account the specifics of both technologies for developing a reliable characterization system to control the IFE target parameters. In the FST-LM, the reconstruction algorithms for tomographic data processing of the target layering stage are used in a hundred-projection microtomograph with a spatial resolution of 1 μm [9, 10]. Moving target tracking is a challenge task and it is becoming increasingly important for IFE applications.

A promising way for *online* measurement of the actual position and quality of the flying target in the reaction chamber is automatic target tracking algorithm based on the Fourier holography [11, 12].

The targets' gravitational loading (one-by-one target injection) from the LM to the TCs (vertical and horizontal) with a rate of 0.1 Hz at T = 5 K is shown in **Figure 4.**

Peculiarities of the FST layering process consist in the following:

1. A fundamental difference of the FST layering from the generally accepted approaches is that it works with free-standing and line-moving targets that allow starting developments on the FST transmission line of repeatable operation.

2. High cooling rates combined with high-melting additives to fuel content (**Figure 4a**) result in creation of a stable ultimate-disordered structure with a high defect density or isotropic medium (ultrafine fuel layers).

3. Additives work as stabilizing agents keeping the grain size stable and, as a consequence, keeping the thermal and mechanical stability of the ultrafine cryogenic layers.

4. For D-T mixture (having the molecular composition: 25% of D_2, 50% of DT molecules, and 25% of T_2), just T_2 is considered as a high-melting additive with respect to D_2 and DT (**Figure 4**).

5. Isotropic ultrafine layers have an adequate thermal and mechanical stability for the fuel layer survivability under target injection and transport through the reaction chamber [1, 15–19].

6. An important parameter is the target lifetime within a temperature interval ΔT, in which a stable ultrafine fuel structure can exist. Our study shows that the fuel doping in the range of η = 0.5–20% (neon, argon, tritium) makes this interval greatest possible, from 4.2 K right up to the temperature of fuel melting [1, 19].

7. Vibrations during FST layering are additional and effective means to meet the demands on the fuel layer formation with inherent survival features. Periodic mechanical impact on the fuel is one more option to a fuel structurization. Therefore, we plan experiments using a classical FST-LM combined with a special vibrator for launching the high-frequency waves in the top part of the LC which in turn will work as a waveguide, maintaining a vibration loading on the moving targets during their layering [1, 22].

2.3 FST layering time for direct-drive high-gain targets

In [15], we proposed a model for rapid fuel layering inside moving, free-standing targets. It is based on solving the Stephen's problem for moving boundaries between the fuel phases (gas, liquid, and solid) and for nonlinear boundary condition onto the outer shell surface. The heat transport through the target is conduction through a small contact area. The computational tools allow one to model the layering time as a function of the target and LC parameters and other experimental conditions. The total layering time is typically less than 15 s (for targets less than 2 mm in size).

At the current stage of research, the FST model was adaptable and scalable for IFE targets (~4 mm). For comparison, in our analysis, we consider several direct-drive target designs for different laser energies E_L:

- OMEGA (E_L = 30 kJ) baseline target [23]: a 0.46-mm-diameter polymer (CH) shell with a 3-μm wall and a fuel layer of 100-μm thick;

- High-power laser energy research (HiPER, E_L = 200 kJ) baseline target (BT): the HiPER-scale targets are of two types [18]. The first one (BT-2) is a 2.094-mm-diameter CH shell with a 3-μm wall. The solid layer thickness is 211 μm. The second (BT-2a) consists of a 2.046-mm-diameter CH shell with a 3-μm wall having fuel-filled CH foam (70 μm) on its inner surface. Onto the foam, there is a 120-μm-thick pure solid fuel;

- Classical high-gain target (CHGT, E_L = 1.3 MJ KrF laser) developed by Bodner and coauthors [24]: "...a new direct-drive target design that has a predicted energy gain of 127 using a 1.3 MJ KrF laser, and a gain of 155 using 3.1 MJ." For 1.3 MJ KrF laser, the target specifications are as follows: the vapor cavity has a 1500-μm radius. The pure D-T (190 μm) fuel is surrounded by an ablator that consists of a CH foam (~10 mg/cm^3) filled with frozen D-T (261 μm). The ablator is surrounded by a 1-μ plastic coating (polystyrene, polyimide, etc.) to contain the D-T fuel. The plastic coating is then surrounded by an overcoat of a thin high-Z material such as gold to withstand the thermal chamber environment.

Figure 5 scales the FST layering time for cryogenic targets of a few millimeters long.

Table 1 presents our new modeling results obtained for CHGT, which is of special interest. The FST layering time (τ_{form}) does not exceed 23 s for D_2 fuel and 30 s for D-T fuel. In order to overcome the gravitational fuel sag to the shell bottom, the FST layering uses a moving target that allows avoiding the difficulties inherent in the cryogenic layer formation in the fixed targets. The shell rotation causes a spread of liquid fuel over the inner shell surface, and under certain conditions (sufficient value of τ_{liquid}), it results in a uniform layer formation. This important effect makes topical to study a dynamical spread of the liquid fuel inside the moving target and to develop numerical models of the process. The obtained results (theoretical and experimental) can be found in [10, 15].

Thus, for dynamical fuel symmetrization in a batch of rolling targets (**Figure 3**), the time of liquid phase existence, τ_{liquid}, is a key parameter and must be sufficient for a cryogenic layer symmetrization. This depends on a temperature T_{in} of the target entry in the LC (initial target temperature before FST layering). Decrease in temperature T_{in} will lead to decrease in the total layering time at the expense of τ_{liquid} (**Table 1**). Therefore, when designing the FST-LM for CHGT, it is essential to manage the value of T_{in} in a way that limits the risks and achieves maximum possible growth of τ_{liquid}.

In near-term plans, we consider the approach based on the FST layering method as a credible path for creating a repeatable operating FST supply system (FST-SS). The first step in this direction is the development of the next-generation FST-LM for high-gain direct-drive targets, which are the shells of ~4 mm in diameter with the shell wall of different designs from compact and porous polymers. The layer thickness is ~200 μm for pure solid fuel and ~300 μm for in-porous solid fuel. Most

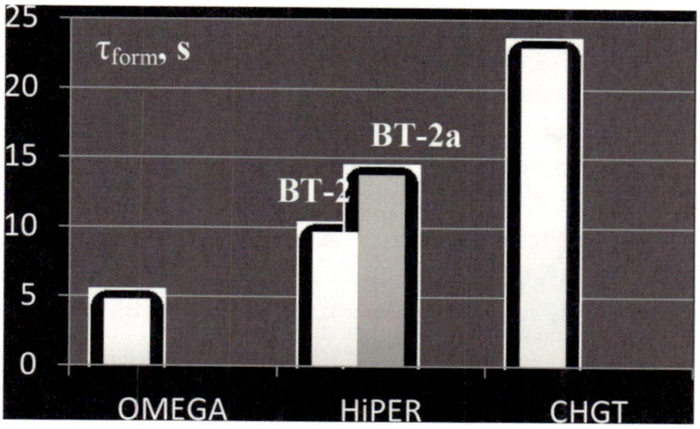

Figure 5.
The FST layering time for several direct-drive target designs was calculated using the computational codes developed at LPI [15].

	D_2 fuel			D-T fuel	
T_{in} (K)	τ_{liquid} (s)	τ_{form} (s)	T_{in}(K)	τ_{liquid} (s)	τ_{form} (s)
35.0	17.48	22.45	37.5	22.14	28.52
27.5	7.08	12.05	28.0	7.87	14.25

Table 1.
FST layering time for CHGT.

of the theoretical efforts have focused on the development of computational models of the IFE target response during FST-formation cycle [1]: fuel filling–fuel layering–target injection. Using the codes allows planning experiments and studying the behavior of these targets in the FST-LM.

Current status of the FST technologies underlies the future research that focuses on the FST-LM prototype development, challenges and advances in IFE target fabrication. We use the CHGT to design a high rep-rate FST-LM and analyze recent experiments with different LCs. Our experiments were made with the mockups of different designs, and the required LC geometry was found. The time-integral performance criterion is that the target residence time t_{res} in the LC must be more than the fuel layering time τ_{form}. **Figure 6** shows three mockups: mockup 1 M-1 (one-fold spiral), $t_{res} = 9.8 \pm 0.5$ s; mockup 2 M-1 (two-fold spiral), $t_{res} = 23.5 \pm 1.7$ s; mockup 3 M-2 (three-fold spiral), $t_{res} = 35.0 \pm 2.0$ s.

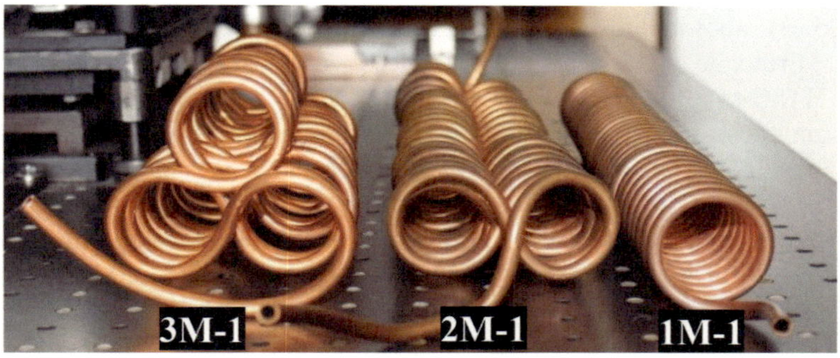

Figure 6.
Different LC mockups which are planned to be used in the FST-LM to promote development of effective affordable technology alternatives.

These measurements show that 4-mm targets can be manufactured by the FST layering method using n-fold-spiral LCs at n = 2, 3, because they maintain the gain in time of the target residence in the LC and in fuel layer symmetrization during target rolling. Note that currently only curved LCs in a specialized geometry and moving targets are successful for developing the FST-LM of repeatable operation, which works with a target batch rolling along the LC.

Our latest effort underlies the future research on creation of the FST-LM as a means of a steady-state target-producing device, which is compatible with a noncontact schedule of the target delivery to the reaction chamber.

3. Target repeatable delivery

During the target delivery, it is necessary to maintain the fuel layer quality in the process of target acceleration and injection. For this reason, the target must be placed into a special target carrier (sabot). Using sabots, there occur some contact problems. Because of a tight seal between the sabot and the barrel, any damage of the barrel and the sabot surface will affect the injector performance and sabot reusing.

Recently, we have started the investigation into magnetic levitation as an alternative technology of noncontact manipulation, positioning and transport of the finished cryogenic targets. From the moment of discovery of bulk high-temperature superconductors (HTSC), which can stably levitate above the permanent magnets, the magnetic levitation (maglev) transport systems are of great interest for their potential application. The transport process with levitation results from the direct use of the diamagnetic characteristics of the HTSC materials. Their unique features can be exploited in the process of levitation and guidance of a HTSC-sabot as well.

In IFE research, this approach attracts a significant interest due to maglev potential for almost frictionless motion. The challenging scientific and technological issues associated with this task are being addressed through a combination of material selections and material property measurements, mathematical and experimental modeling, demonstration of the HTSC-sabot acceleration in laboratory-scale tests.

3.1 Noncontact acceleration system

A noncontact acceleration system proposed at the LPI is a combination of three basic elements: (1) electromagnetic acceleration system (EM-AC), which includes the field coils generating the traveling magnetic waves, (2) levitation system (permanent magnet guideway or PMG), which includes a magnetic rail (or magnetic track), and (3) sabot including several HTSC components. **Figure** 7 illustrates the operational principle of the system. During acceleration, the target is protected with a levitating HTSC-sabot, and the diameter of the barrel exceeds the sabot diameter. This is a small-scale prototype under construction of a hybrid accelerator "EM-AC- + PMG" at the LPI. The concept of "EM-AC + PMG" is completed and the proof-of-principle (POP) experiments have confirmed the benefits of this approach.

The prove-of-principle (POP) experiments (**Figure 8**) on magnetic acceleration of the levitating HTSC-sabot are made in the mutually normal magnetic fields: the first is **B**1 (from the field coil to move the HTSC-sabot) directed along the acceleration length and the second is **B**2 (from the permanent magnets to counteract the gravity) directed normally to the acceleration length (**Figure 7c**). The Meissner effect [25] dictates that both magnetic fields generate the surface currents around the superconductor (in our case, it is the HTSC-sabot) in corresponding

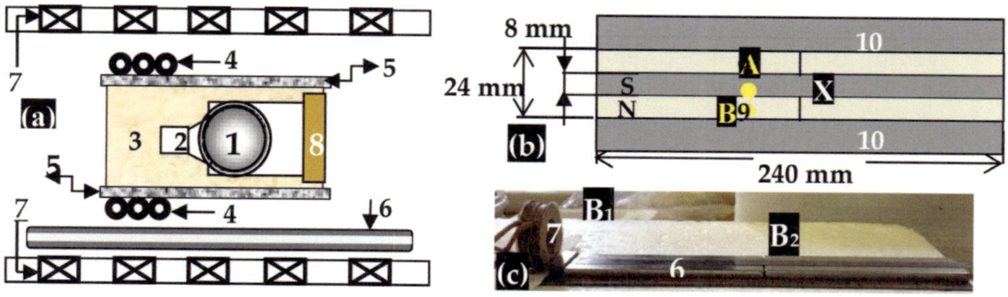

Figure 7.
*Schematic of "EM-AC + PMG". In (a): 1—target, 2—conical support of the target nest, 3—sabot matrix (polymer), 4—HTSC-coils (driving body based on superconducting MgB2-cables), 5—HTSC-plates for providing a stable levitation along the magnetic track, 6—magnetic track, 7—field coils, 8—protective cover; in (b) and (c): parameters and a side view of the experimental setup with directions of the driving force **B1** and the levitation force **B1** (9—ferromagnetics, 10—permanent magnets).*

directions that leads to its simultaneous acceleration and levitation (simultaneous presence of the driving force along the vector **B1** and of the levitation force along the vector **B2**).

The experiments were conducted under the following conditions.

- To ensure the successful acceleration, the field coil and HTSC-sabot are fixed over the magnetic track of the PMG-system so that the horizontal axis of the coil and the HTSC-sabot coincide (**Figure 8b**).
- The HTSC-sabot motion has been driven by the electromagnetic pulse generated by the field coil, that is, by using a running gradient of the magnetic induction (accelerating running pulse, or ARP).
- As the HTSCs are diamagnetic, the HTSC-sabot is pushed out from the area of a stronger magnetic field.
- The starting parameters of the coil pulse: the pulse duration is 1 ms, the current amplitude is 200 A, the maximum magnetic induction is 0.35 T.
- The temperature in the experiments was T = 80 K for the following reason. Our previous studies have demonstrated a high efficiency of "HTSC-PMG" interaction in a wide temperature range ΔT = 5–80 K [4]. Consequently, it is possible to study the HTSC-sabot acceleration at temperatures close to 80 K, that is, under the nitrogen cryogenics. This is especially important because it

Figure 8.
HTSC-sabot friction-free acceleration by the driving electromagnetic pulse generated by the field coil (T = 80 K). In (a): HTSC-sabot before acceleration; in (b) HTSC-sabot starts motion, in (c) and (d): in-time development of the acceleration process (the gap between the HTSC-sabot and the magnetic track is keeping unvarying with time).

allows the acceleration experiments to be feasible at T~80 K taking into account that such experiments at T~18 K are unpractical inside a small test chamber of the cryostat.

Figure 8 shows the results of the demonstration tests: a set of freeze-frame shots of the acceleration process of the levitating HTSC-sabot at T = 80 K using the linear PMG-system.

The HTSC-sabot is trapped (**Figure 8a**) and accelerated in front of a magnetic traveling wave (**Figure 8b–d**). It reaches a velocity of 1 m/s and keeps this velocity on all sabot-track length of 22.5 cm (motion time is t = 0.22 s). The levitating drift is not observed. Technologically, this allows a convenient spacing of the multiple coils (also called a multiple-stage accelerator, **Figure 7a**), and leads to realizing very high velocities of the HTSC-sabot. It is necessary to highlight that in the experiments, we did not use a special driving body, and the obtained result of 1 m/s is due to the surface currents in the HTSC material itself arising from the magnetic field **B1** generated by the field coil.

Thus, a friction-free HTSC-sabot transport can be realized with the levitation devices using superconductors and permanent magnets. The continuous space range of the stable position of the levitated HTSC-sabot has been demonstrated by the experimental results. Features of the device concepts and their future applications in the noncontact delivery system are discussed below.

3.2 HTSC materials

Generally, superconducting material selection for sabot manufacturing is defined, first of all, by the temperature requirement for IFE targets which must be at T = 18.3 K [1–3]. Superconductors are classified into two types [25], called type-I and type-II, based on their diamagnetic properties (magnetic susceptibility $\chi < 0$). Type-I superconductors (low-temperature superconductors) are in the state which is called "perfect diamagnetism" (the Meissner effect at which the magnetic lines bend around the superconductors). As the applied magnetic field increases, so does the opposing magnetization until the field reaches the critical field B_C, whereupon the superconductivity disappears. Since the type-I superconductors have the critical temperature $T_C < 10$ K (i.e., their heating above 10 K destroys their superconductivity), they cannot be considered as candidates when developing a maglev transport system for application to IFE. Besides, type-I superconductors typically have the critical field values too low for practical applications.

In the study, we use the samples from HTSCs, which are known to be type-II superconductors. They have two values of the critical magnetic field B_C (called first B_{C1} and second B_{C2}). Below B_{C1}, type-II behaves similar to the type-I. When the applied magnetic field is between B_{C1} and B_{C2}, the magnetic field penetrates the type-II superconductors in the form of quantized magnetic flux lines (either tube or vortices), and they become a mixture of the normal and superconducting states. Emphasize that inside each magnetic flux tube, superconductivity is locally destroyed. Such materials can be subjected to much higher external magnetic fields and remain superconducting. This property is used for obtaining strong magnetic fields under the conditions of no thermal losses when the high currents are passing through HTSCs. In addition, the HTSC materials with critical temperatures in the range of 90–120 K have received a great deal of attention because they can be maintained in the superconducting state with liquid nitrogen (77 K).

The second issue under superconducting material selection is structural characteristics of HTSCs, which influence on the potentialities of their levitation properties. The authors of [13] revealed many interesting features related to this problem.

Different HTSC materials can have different microstructures, from so-called "superconducting glass" (superconducting ceramics) to microstructures like a type of "mosaic" with macro-, meso-, and microlevels of material ordering. This creates favorable conditions for obtaining an optimal HTSC microstructure just for taking into account the design specifics of a noncontact delivery system intended for the IFE experiments.

At the same time, it is necessary to note one more important feature related to type-II superconductors. It is a flux (or vortex) pinning. Vortex pinning results from spatial imperfections (or defects) in the material that produces local reductions of the free energy of a flux line [25], thus attracting and holding vortices to these locations. In many respects, the basic magneto-mechanical phenomenon responsible for levitation is a result of the magnetic flux pinning inherent in the interaction between a magnet and a type-II superconductor. In the mixed state, the flux lines interact with different defects and may become pinned to them (frozen in the bulk superconductor). Such defects (e.g., crystal lattice defects, grain boundaries, twin planes, stacking faults, etc.) always exist in real superconducting materials. They could work as pinning centers (including pinning by surface roughness or at a step-like surface relief), avoiding the vortex motion and consequently the energy dissipation. A vortex state looks like a "frozen" in the superconducting material, and any spatial motion of the superconductor will cause the flux tubes to move. In order to prevent that, the superconductor remains "trapped" in its original state (be it levitating at the fixed point or under motion along a magnetic track).

Thus, the bulk of type-II, HTSC materials breaks down into two regions: superconductive—from which the external field is completely expelled, and normal—through which the external field penetrates. The diamagnetic characteristics of the material are more or less pronounced depending on its "degree of superconductivity." These features can be exploited in the process of HTSC-sabot levitation and guidance. Therefore, the superconducting material science and technology is of critical importance. Currently, there are many structural efforts to enhance the pinning properties in HTSCs by creating structural defects in them using different techniques. These structural defects can be in the form of periodic pinning arrays or random pinning distributions (i.e., with different ordering states of a vortex lattice) to improve the maglev properties of the HTSCs.

In our experiments, the HTSCs are superconducting ceramics based on $YBa_2Cu_3O_{7-x}$ (or Y123; production of LPI) and superconducting tapes of second generation (2G HTSC) based on $GdBa_2Cu_3O_{7-x}$ (or Gd123; production of SuperOx, Ltd.). The obtained results have shown that these HTSCs can be successfully used to maintain a friction-free motion of the HTSC-sabots, and also to provide a required stability of the levitation height over the whole acceleration length due to the pinning effect. This becomes more viable because the critical temperature of Gd123 and Y123 is Tc $\sim$ 90 K, which is nearing the boiling point of nitrogen (**Table 2**).

The HTSC-sabot designs were in the form of an "open parallelepiped" (its cross section forms a trough) or in the form of a "hollow parallelepiped" (its cross section forms a square).

HTSCs	Density (ρ, g/cm^3)	B_C at 0 K (T)	T_C (K)
Y123 [4, 6]	4. 33	>45	91
Gd123 [5, 26]	3.25	> 45	92

Table 2.
Parameters of the HTSC materials used in the experiments.

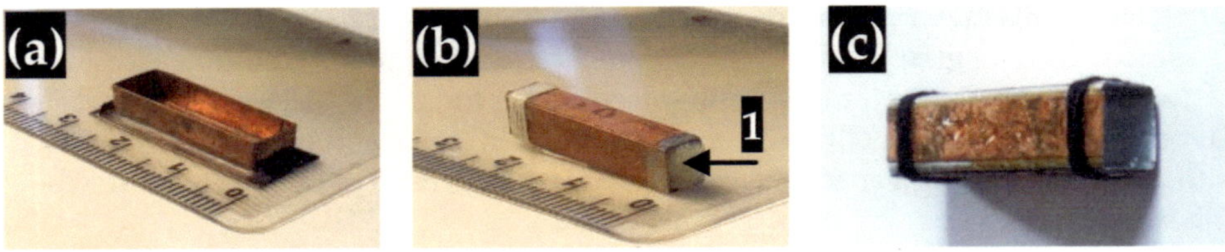

Figure 9.
Usual HTSC-sabot designs used in the experiments. In (a): "open parallelepiped" (model#1); in (b): "hollow parallelepiped" (model#2 + polymer foam (1)); in (c): model#3—"hollow parallelepiped" (empty).

Here, we set the task of accelerating different HTSC-sabots (**Figure 9**) over different PMG-systems to study the stability of the main levitation parameters: the load capacity (mass of an object which HTSC-sabot can maintain), the space locking (three-dimensional stability of HTSC-sabot), and the gap between the HTSC-sabot and the magnetic track (levitating drift). The HTSC-sabot parameters are as follows:

- Model #1, "open parallelepiped": Gd123 tape thickness is 0.5 mm, length inside is 25 mm, width inside is 8 mm, height is 4 mm, total mass is 1.26 g;
- Model #2, "hollow parallelepiped": Gd123 tape thickness is 0.3 mm, internal sizes are $4 \times 4 \times 24$ mm^3, total mass is 0.97 g (together with filling of polymer foam which mass is 0.38 g);
- Model #3, "hollow parallelepiped": Gd123 tape thickness is 0.3 mm, internal sizes are $4 \times 4 \times 30$ mm^3, total mass is 0.59 g (no polymer filling).

In the experiments, the force F driving the diamagnetic Gd123-tapes is given by [27]:

$$F = \frac{\chi}{2\mu_0} V_S \frac{dB_x^2}{dx}, \tag{3}$$

where μ_0 is the permeability of vacuum, V_S is the superconductor volume, x is the acceleration axis, **B** is the magnetic induction produced by the field coil (in our case **B**1). Since for diamagnetic $\chi < 0$, then this force is contrariwise to the gradient of the magnetic induction in the x-direction. Therefore, the HTSC-sabot is pushed out from the area of a stronger magnetic field that defines its behavior in the PMG-systems.

3.3 PMG-systems

The characteristics of the permanent magnets composing the PMGs are very important for their performance in terms of levitation force and stability. The PMG-systems of different configuration were used in the experiments (**Figures 8, 11–14**). Our goal was to demonstrate not only the levitation stability of the HTSC-sabots, but also their transport over the PMGs with a guidance force resulting in either linear or circular motion in a "tight space."

The PMGs [4, 5] were constructed on the basis of neodymium permanent magnets with an axial magnetization (MIDORA, Ltd.). A magnetometer with a sensitivity of 280 mV/T was applied with the following performance data: the positioning accuracy of a sensitive element is 0.1 mm, the measuring range is +1 T, and the absolute error is +0.005 T. The maximum magnetic induction (0.33 T right at the track's surface) was for the linear PMG (**Figure 11**) along the acceleration direction OX for a magnetic track of 24-cm long.

Note that the PMG optimization is the most critical problem of practical interest since it serves as a continuous magnetic track to generate the required magnetic field by rare-earth permanent magnets (made from an alloy of neodymium, iron, and boron to form the $Nd_2Fe_{14}B$) and inserts of soft magnetic (ferromagnetic). Therefore, a feasibility study of the key technical issues such as influence of the PMG-fields of different configurations on the mechanical and timing performances of the HTSC transport process and active guidance due to different driving pulses (mechanical and electromagnetic) are under way.

In this chapter, we present an analysis of dynamic behavior for two proposed PMGs consisting of different arrangements of the permanent magnets (different shape and size). The levitation experiments in specifically designed configurations (**Figure 10**) with strongly pinned superconductors (Gd123 and Y123) display a repeatable PMG operation, allowing a simultaneous demonstration of linear and lateral stability. We have also studied the issues of how the geometrical and loading characteristics of the HTSC-sabots can affect their levitation capability at different constraints of the PMG cross section.

The main idea in the PMG construction is that the magnetic track must allow the HTSC-sabots to move freely only in one direction in order to avoid any contact with a stronger magnetic field, which pushes out them and return them to their initial trajectory (according to the Eq. (3)). This is due to the fact that any spatial motion of the HTSCs will cause the magnetic flux tubes to move. A reasonable plan to prevent this effect in the lateral direction is to made the linear PMG with the magnet poles aligned antiparallel to each other (N-S-N) for producing a considerably strong gradient for a side-to-side motion (**Figures 7, 8, 10a**). The circular PMG consists of a disk of NdFeB permanent magnet (OD = 100 mm, ID = 50 mm, 5 mm thick) embedded in the soft ferromagnetic holder to realize a required distribution of the magnetic field along the width of the magnetic track (**Figure 10b**). Since the flux tubes are magnetic fields frozen in the superconducting material, the very superconducting material itself creates a force to inhibit any motion in relation to the magnetic field, and the HTSC-sabot remains "trapped" in its trajectory. This is an efficient scheme for optimizing the levitation and guidance forces which is considered as a base to perform the search of an optimal PMG.

We have found that not only the linear PMG-systems (**Figures 8** and **10a**) but also the circular ones are promising candidates aimed at development of HTSC-maglev transport system for high-velocity target applications, target trajectory correction, and creation of a precise injector. The circular PMG testing under a typical optimization of the levitation stability has proven its robustness and efficiency. **Figure 11** demonstrates the HTSC-sabot motion with different velocities over the circular PMG shown in **Figure 10b**. In the experiments, we use a mechanical

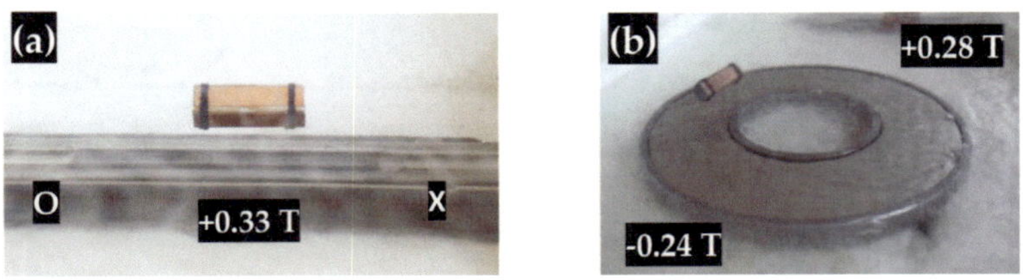

Figure 10.
*Acceleration of the HTSC-sabot (model #3, **Figure 9c**) over the linear (a) and circular (b) PMGs at T = 80 K.*

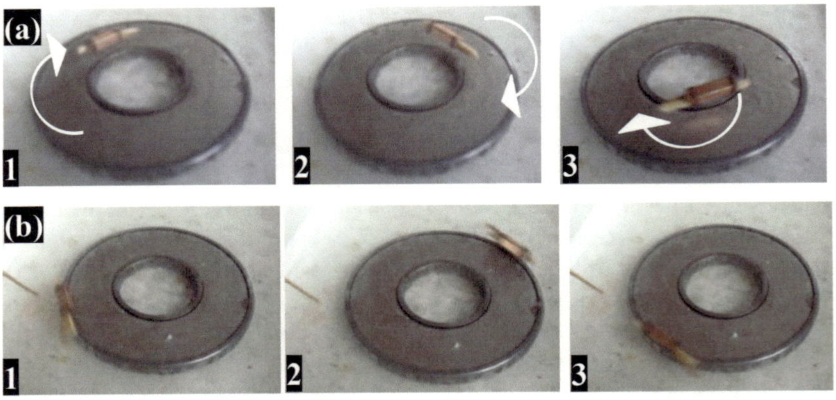

Figure 11.
Different velocities of the HTSC-sabot (model #3, load capacity: a piece of wood) over the circular PMG at T = 80 K. In (a): internal orbit; in (b): external orbit.

driving pulse to simply change the initial conditions related to placing the HTSC-sabot on different trajectories. The findings accessible from these experiments are the bases to provide the conditions when the HTSC-sabot can be accelerated in a stable orbit under different load capacities, which is directly related to the safe operation and design of the whole system. Below, we present the experimental results obtained in this area. **Figures 12** and **13** show a stable acceleration of HTSC-sabot (Model #1) over the linear and circular PMGs with several samples surrogating both spherical and cylindrical targets.

In both cases, after being disturbed during motion, the HTSC-sabot has a disturbance-induced velocity that, in general, can result in changes in the levitation and guidance forces and can cause some serious malfunctions. If these induced

Figure 12.
Acceleration of the HTSC-sabot over a linear PMG at T = 80 K. In (a): HTSC-sabot moves along the magnetic track; in (b): 1—HTSC-sabot is on the middle of the track, 2—HTSC-sabot exit out of the track (load capacity: 5 spherical targets of 0.6 mg each).

Figure 13.
Acceleration of the HTSC-sabot over the circular PMG at T = 80 K. In (a): before acceleration; in (b) and (c): during acceleration (load capacity: three cylindrical targets of 1.1 g each); in (d): before acceleration; in (e) and (f): during acceleration (load capacity: two cylindrical targets of 1.1 g each are placed in the same HTSC-sabot in such a way that one of them is vertical, and the other is horizontal).

external disturbances are considerable, vertical and lateral displacements of the levitating body may occur simultaneously. In this connection, it was very important to study the levitation stability of the HTSC-sabots.

Our findings (**Figures 8, 10–13**) have shown that a basic phenomenon responsible for the levitation stability is the flux pinning effect inherent in the interaction between a type-II superconductor and a permanent magnet. For specifically designed configurations of PMGs (peculiar distribution of the magnetic field), the flux-pinning is tending to enhance the stability of HTSC-sabot levitation, and strongly pinned superconductors (Gd123) display high stability, allowing the demonstration of striking effects, such as vertical, lateral, or inverted levitation. They look like they are pinned to a magnetic track so they can stably levitate over permanent magnets without any active control. Thus, the HTSCs can be designed to enhance the effect called "enhanced flux pinning." It is of a great importance for target trajectory correction during its delivery inside the assembly of "HTSC-sabot + target." Several remarks should be made here.

An idea of using the magnetic field to control the "HTSC-sabot + target" transport is very attractive for the following reasons. In HTSC-maglev, the stable transport with levitation is caused by a combination of the Meissner effect (quantum levitation) and the flux pinning in HTSCs (quantum locking).

As it is mentioned above, in HTSCs the magnetic field is not excluded completely, and the superconductor tries to keep the magnetic flux or vortexes pinned in weak areas (e.g., grain boundaries or other defects). Energetically, this means that the vortexes favor to be located in the bulk of HTSCs where defects exist. Any spatial displacement of the HTSCs causes the magnetic flux motion. Using the vortex physics [26] under the PMG construction, we have succeeded in controlling the magnetic field so that there are some directions where the HTSC-sabots can move, and there are some directions where the HTSC-sabots remain "trapped" (located in a "tight space").

Along the magnetic track (or acceleration length, both linear and circular), there are no magnetic field changes, which allows the HTSC-sabots to move forth and back with no energy loss. Normal to the acceleration length, the magnet poles are aligned antiparallel to each other (N-S-N, see, for example, **Figure 7b**) that produces a considerably strong gradient along the width of the magnetic track. This gradient prevents the motion of HTSC-sabots, and they remain located in the transverse direction. In other words: (1) flux-pinning makes the HTSC-sabot motion trapped in the space within a PMG field; (2) flux-pinning makes the HTSC-sabot orientation fixed in the space so that they will not reorient themselves without any external action (so-called a three-dimensional locking of type-II superconductors). This process of locking by height and orientation reduces any undesirable wobble during HTSC-sabot movement. Thus, the obtained results indicate that we have an effective set of tools (quantum levitation and quantum locking) for a noncontact acceleration of the HTSC-sabots in the mutually normal magnetic fields generated by the field coil and the PMG-system.

3.4 HTSC-sabot design

The HTSC-sabot design is a vital point in the process of its transport with levitation. The most striking examples are the HTSC-sabots, the shape of which corresponds to Model #4 and Model #5 (**Figure 14**):

- Model #4, "hollow parallelepiped + 5 wings": Gd123 tape thickness is 0.3 mm, Model #2 ($4 \times 4 \times 30$ mm^3) + 5 wings (12×12 mm^2), total mass is 1.46 g.

- Model #5, "hollow parallelepiped + 5 wings" on a Gd123 tape ($35 \times 12 \times 0.3$ mm^3): total mass is 1.97 g.

Nevertheless, **Figure 15a** shows that the Model #4 (the Model #2 in assembly with five wings) as an independent target carrier is inefficient. The wings keep the Model #4 "nonlevitated" so that it comes into contact with the magnetic track. However, if using the same Model #5 (the Model #4 placed on a superconducting Gd123 tape), the levitation effect occurs again (**Figure 15b**), because the Model #5 is a combination of the levitating Gd123 tape and the Model #4 as a load capacity.

These results can be explained by a special mapping of the magnetic lines which are bending around the Model #4 creating the magnetic field close to the second critical value, B_{C2}, or even more this value (T = 80 K is close to TC = 92 K). At a step-like surface relief (critical bending of the magnetic lines), the magnetic field is able to considerably slip through the HTSC material of the Model #4, and the normal cores of vortexes begin to adjoin and then the volume superconductivity disappears. In other words, under roughening of a surface, the number of vortices becomes so numerous that there is no space left for superconductivity, and the superconducting material becomes less and less diamagnetic.

3.5 HTSC-driving body

The driving body is a component of the HTSC-sabot (**Figure 7**). In our study, we consider a superconductor on the basis of MgB_2-cables because of its potential for applications at high magnetic field [28, 29]. However, under using MgB_2-driving

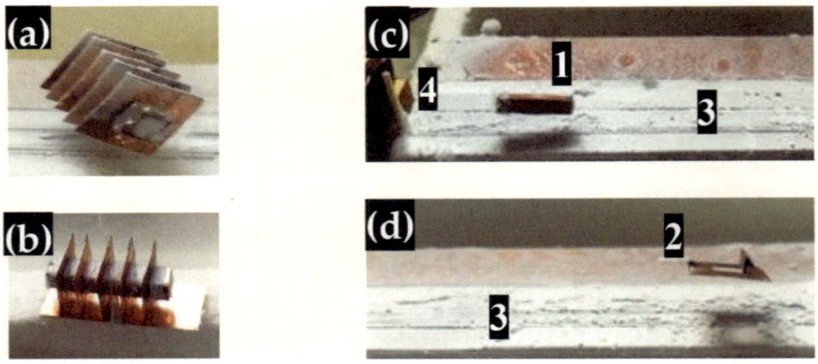

Figure 14.
HTSC-sabots with several detached wings. In (a) and (b): overview of model #4 and model #5; in (c) and (d): acceleration (T = 80 K) of the different components of model #4 (1—hollow parallelepiped, 2—wing) along the magnetic track 3 by the field coil 4.

Figure 15.
Levitation of the model #4 and model #5 at T = 80 K. In (a): lack of levitation for the model #4 (its motion is caused by a magnetic track inclination, that is, by gravity under the inclination angle less than 10°); in (b): the same HTSC-sabot (model #4) levitates just on a Gd123 tape (i.e., model #5 levitates stably).

body, the acceleration parameters which are of interest for IFE (injection velocity V_{inj} = 200 m/s) become unsuitable for the laboratory-scale tests. Therefore, the POP experiments (**Figure 8**) have been carried out with HTSC-sabots (Gd123) without any MgB_2-coils in their design. The HTSC-sabot (Model #2) is accelerated using the magnetic field **B1** generated by the field coil (ARP). The acceleration process maintenance is caused by the surface currents induced in the bulk Gd123 itself due to ARP, which results in arising the driving force along the acceleration length. The HTSC-sabot obtains a velocity of 1 m/s and keeps it over the whole magnetic track (22.5 cm). This is a demonstration of the one-stage accelerator.

Below, we discuss the issue related to a multiple-stage accelerator. The first problem is as follows: what characteristics of MgB_2-cables are required to reach the required lower limit on the injection velocity V_{inj} = 200 m/s. We list below some distinguish features of MgB_2, which are important for our study [28, 29]:

- MgB_2 critical temperature is TC = 39 K, which is twice above than for Nb_3Sn, and four times above than for Nb-Ti;
- MgB_2 is a promising superconductor for applications in the temperature range 15–20 K which meets the temperature tolerance for the IFE targets which must be at T = 18.3 K before the laser shot to obtain the maximum energy yield from the fusion reaction [1–3];
- MgB_2 possesses high values of the critical current at a rather small sensitivity to intergranular contacts;
- simple chemical composition and low cost of the initial components for its synthesis;
- achievement of typical magnetic fields more than 1.5–2.0 T at lower capital investments and at lower cost of commercial operation;
- stability of MgB_2 characteristics under the conditions of radiative environment;
- due to a weak anisotropy of the critical properties, the MgB_2-cables can be well-shaped that is of a great importance for optimizing the current density in superconducting coils. Besides, the MgB_2-cables can be of round or rectangular cross section and have a small weight. These are most important parameters for practical applications of MgB_2-cables in the HTSC-sabot design.

For estimations of the acceleration length, La, for a multiple-stage accelerator with a superconducting driving body (in our case MgB_2-cables), we use the following ration [8]:

$$L_a = \frac{\pi}{2} \cdot V_{inj}^2 \cdot \frac{R_{FC}}{R_{SC}} \cdot \frac{M_{sab}}{F_{pin} V_S}, \ F_{pin} = J_C(B_0, T_S) \cdot B_0, \tag{4}$$

where M_{sab} is the mass of the "HTSC-sabot + target" assembly, R_{FC} is the field coil radius, R_{SC} is the radius of the superconducting coils (R_{FC}/R_{SC} = 5), F_{pin} is the pinning force density, J_C is the critical current density, which depends on the magnetic induction in the coil center B_0 and superconductor temperature T_S. The value of J_C (defined as the current density where the pinning force and the Lorentz force become equal) determines the onset of resistivity [28–30]. In (4), a difficulty arises in calculation of L_a because only knowing the relationship between J_C and B_0,

External field (T) [30]	1	0.5	0.25
Critical current (A) [30]	2500	4000	5000
Sabot acceleration (g)	800	640	400
Acceleration length (m)	2.5	3.125	5.0

Table 3.
MgB_2-driving body acceleration efficiency at T_S = 20 K.

the pinning force density F_{pin} can be found for the superconducting coils proposed for the sabot acceleration. For MgB_2-cables of 1.18 mm in diameter, the critical current vs. the magnetic field at temperatures of 4.2, 9.8, 15, 20, and 25 K was measured in [30]. Using these data, we have made the calculations under the actual operating conditions: (a) the target design is CHGT, and its mass is 5 mg (see Section 2.2); (b) the HTSC-sabot is "open parallelepiped" to exclude a bend of the Gd123 tapes (see Section 3.2); and (c) the mass of the assembly "HTSC-sabot-+ CHGT" is 0.5 g. The calculation results are presented in **Table 3**.

Thus, using the MgB_2-driving body allows not only to accelerate the reactor-scaled targets to the required injection velocities, but also to provide the system performance without exceeding the acceleration limits at 500 g. As one can see from the **Table 3**, the MgB_2 coils (with parameters $2\pi R_{SC}$ = 24 mm, B_0 = 0.25 T, J_C = 5000 A) yield V_{inj} = 200 m/s at 400 g on the acceleration length of 5 m.

Note that several important aspects related to a practical engineering are as follows:

- In our study, we have proposed the PMG configuration allowing in-space equilibrium position of the HTSC-sabot during its acceleration (it goes along a whole magnetic track with the same levitation height and orientation).

- Taking into account that experimentally the HTSC-sabots keep their speed after acceleration pulse, they can be extra accelerated by using a multiple-stage accelerator.

- Superconducting cables can be used not only in the driving body but also in the field coils. If these coils carry a current, which is less than the critical current, then large magnetic fields can be generated without heat generation.

- In our model, the MgB_2-driving body represents a magnetic dipole (MD). The MD acceleration is carried out by a traveling magnetic wave or ARP at the consecutive switch of the field coils. From the view point of a relative positioning of the ARP and MD, the steady case is realized when the ARP pushes the MD but does not pull it for itself, that is, the area of a phase (longitudinal) stability is on a forward slope of the ARP [31]. In the accelerating equipment, it is referred to as a principle of automatic phasing. This principle will be inherent for the MgB_2-driving body, because, as a superconductor, it will be pushed out from the area of a stronger magnetic field.

Especially, note that the injection velocities $V_{inj} \geq 200$ m/s are not a problem for the proposed noncontact schedule of the target delivery. It can successfully be used in creation of a hybrid accelerator for future IFE power plants.

4. Target protection system

Target delivery into IFE power plant requires target acceleration (accompanied with mechanical and thermal loads) and repeatable injection into the reaction chamber (additional thermal loads). For this reason, the problem of using the cryogenic targets in the IFE experiments or in a future reactor includes not only an issue of fabricating the qualitative cryogenic fuel layer (nonuniformity <1%, roughness <1 μ), but also an issue of target delivery at the laser focus under conditions of the layer parameter survival. In our study, a number of protection techniques have been proposed and examined with the aim of risk minimization in the process of target acceleration and injection.

A promising direction for survivability of fuel layers is application of external target protective coatings, which reduce the risks of the fuel damage under the radiation exposure from the hot walls of the reaction chamber: cryogenic coatings (from the solid D_2, H_2, or Xe), metal coatings from Au, Pt, Pd, and their alloys, application of a double protection: "metal + cryogenic" (**Figure 16a**). Below, we demonstrate the practical possibilities of this direction.

To obtain the results in **Figure 16b**, an additional procedure is added in the FST formation cycle. First, the metal coating made from Pt/Pd is deposited on the CH-shell. Then, the shell is filled with the D_2 gaseous fuel with 3% Ne as doping agents. The next step is D_2 layer fabrication by the FST layering method.

The sabot is also a special element of the target protection system [12]. An important feature of its design is the shape of a target nest. Our study shows that a proper choice of the nest shape makes it possible to significantly increase the upper limit of the permissible overloads and to minimize the injector size. Being based on the discrete-continuous physical model of the shell stress, a simulation code SPHERA is developed that makes it possible to define the stress and deformation arising in the target during the acceleration. A shape analysis of the sabot bottom (in the target nest area) during the target acceleration is carried out for three sufficiently different cases: (1) flat bottom, (2) semispherical bottom with $R_n > R_t$ (R_n and R_t are the nest and the target radii, respectively), and (3) conical bottom (**Figure 7**). Important conclusions followed from these calculations are listed below:

- Permissible target overloads for a flat bottom are 50 times smaller than those for semispherical supports with clearance less than 5 μm; at clearance higher than 20 μm, the stresses growing in the target are close to those of the flat bottom.

- Use of a conical bottom with the angle in cone basement equal to 87° provides a 20-times increase of the permissible overloads; technologically, the conical bottom has much better predictive estimate than the semispherical one.

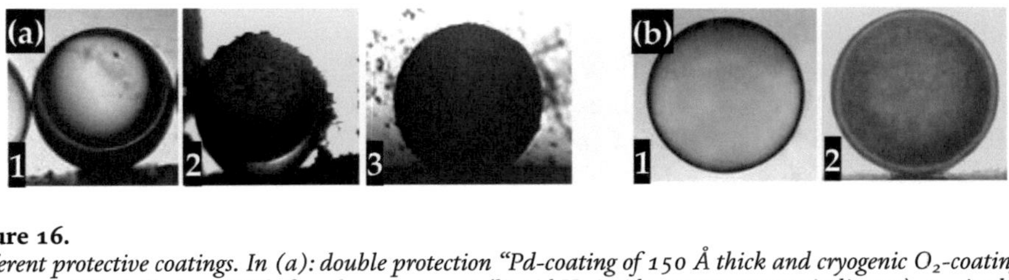

Figure 16.
Different protective coatings. In (a): double protection "Pd-coating of 150 Å thick and cryogenic O_2-coating": 1—1.2-mm CH-shell at 14.6 K before the experiment (liquid H_2 inside as temperature indicator), 2—in the top part of the shell (from the outside), there is a solid deposit of oxygen (T_{tp} = 54.3 K), 3—after operation of the piezoelectric vibrator [22], the oxygen snow becomes redistributed onto the outer shell surface; in (b): single protection "Pt/Pd-coating of 200 Å thick": 1—1.5-mm CH-shell before the experiment, 2—cryogenic target at 5.0 K with a uniform D_2 layer of 50 μm fabricated by the FST-layering method.

The next step is a shield (or cover) for application to protect injected target from a head wind of a residual gas. It has been considered since 1982 [32]. In [16], we have proposed a new design of a protective cover made from solid xenon or deuterium. At the current research stage, we have analyzed the cover and the target interaction with the reactor chamber environment using the direct simulation Monte Carlo (DSMC) approach as well as using the results of numerical studies of gas flow interaction with bodies. The following relations were considered:

- motion equation

$$L = Ut - F_D t^2 / 2M_b, \tag{5}$$

- velocity in a laminar circular wake behind the cover [33]:

$$u(x,y) = U[(\pi C_D)/32] \cdot [(2R/x)RF_D] \exp\left(-\eta^2\right), \tag{6}$$

- drag force equation for a sphere [34]:

$$\mathrm{F}_D = -(3/4)a^2 nu\left(m_g k_B T/2\pi\right)^{1/2} F_1 \mathrm{k}. \tag{7}$$

Here $F_1 = F_{1o} + F_{1p}$, $F_{1o} = 8\pi(8 + \pi)/9$, $F_{1p} = -21.28/\xi^2$ (for two equal spheres), **k** is the unit vector in a target motion direction, L is the distance between the body (target or cover) and the burn area, R is the characteristic dimension (target or cover), t is the in-flight time, $\mathbf{F_D}$ is the drag force, M_b is the mass (target or cover), U is the injection velocity, u(x, y) is current velocity, η and ν are the dynamic and kinematic viscosities of a residual gas, M is the Mach number, m_g is the mass of residual gas, a is the sound speed in gas, k_B is the Boltzmann constant, F_{1o} and F_{1p} are the dimensionless drag coefficients, ξ is the dimensionless distance measured in target radii, and C_D is the coefficient of gas molecule accommodation by the target surface. The following parameters are used in our estimations: velocity is 250 m/s, residual gas is Xe at 0.5 Torr, reactor chamber radius is 5 m, cylindrical cover from solid Xe with a mass of 87 mg, and target mass is 5 mg (CHGT).

In the drag force estimations, two cases were considered: solitary and joint flight of target and cover. Correction for the solitary case (effect of wake) is about 30%. The estimations show that due to the drag force action, the distance between the target and the cover rises from the initial 1 mm at the moment of injection up to 15 mm at the center of reaction chamber. Thus, the drag force provides necessary separation of the cover and target inside the reaction chamber. The protective cover forms a wake region (**Figure 17a**) with reduced flow velocity and temperature and effectively reduces the gas heat flow by a factor of 4÷5, which is in a good agreement with calculations in [35]. Thus, the concept of protecting the direct drive target in the reactor chamber by a cover moving ahead can be considered as a possible way of solving the target delivery problem.

Note that the problem of target survival is the more difficult the higher the target temperature at the moment of injection. Estimations show that radiation heat flow from the chamber wall is an order of magnitude higher than the gas heat transfer (**Figure 17b**). Therefore, target injection at 5 K is more preferable than at 17 K. This can be realized only with an ultrafine fuel structure [1] because the single crystalline fuel reveals undesirable roughness on target cooling [2].

DOI: http://dx.doi.org/10.5772/intechopen.81518

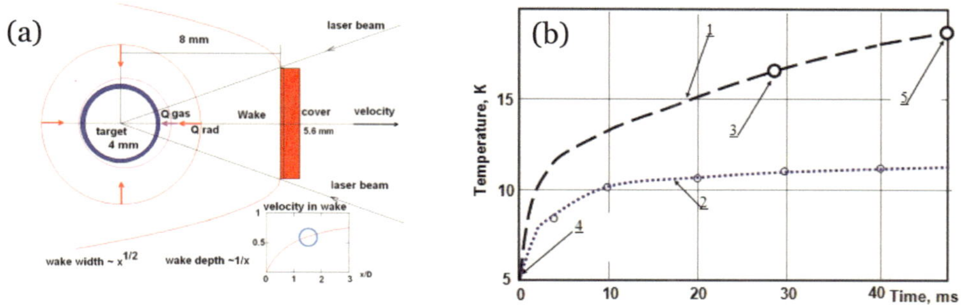

Figure 17.
Protective cover: (a) protective cover forms a wake area in the fill gas to protect a target from the head wind and to avoid the convective heating; (b) target thermal history in IFE chamber under exposure of the wall radiation and gas convection: 4-mm targets with D_2 fuel (200 μm), reactor wall temperature is 1773 K, residual gas Xe − 0.5 Torr (1—target without reflecting layer, 2—target with reflecting layer, 3—injection at 17 K, 4—injection at 5 K, 5—18.5 K, point of destination).

Basing on these results as well as on the results presented in Sections 2 and 3, we propose a multiple target protection system for the effective delivery of a cryogenic target without its damage.

1. Cryogenic layer formation with an isotropic ultrafine fuel structure (which can be referred to as layers with inherent survival features) to reduce the target sensitivity to the external thermal and mechanical loads.

2. Use of friction-free acceleration of the "HTSC-sabot + target" assembly to reduce the heat flux on the target under development of a noncontact delivery system with linear or circular accelerators.

3. Use of conical supports for a target nest in the sabot to reduce the mechanical loads during acceleration of the "HTSC-sabot + target" assembly.

4. Use of outer coatings (cryogenic, metal) in the target design to reduce risks of cryogenic layer damage as a result of target heating by thermal radiation of the hot chamber walls.

5. Coinjection of a target and a protective cover from frozen gases (D_2, Xe) to reduce risks of cryogenic layer damage as a result of target heating by hot residual gases in the reaction chamber.

The important remaining factors of the research include issues of complex technology optimization and system integration.

5. Conclusion

The purpose of this work was to study a repeatable target production and methods of their noncontact delivery in accordance with the scope of MM-IFE program. Various physics and technology problems accompanying IFE target-fueling development were considered, and approaches to their solution were proposed and experimentally tested partially. Our thermal, mechanical, and levitation modeling (theoretical and experimental) are important tools in planning future experiments on MM-IFE and studying IFE reactor fueling.

A particular interest from scientific and technological points of view in IFE progress deals with a cryogenic target that could be delivered to the center of reaction chamber at significant rates. Therefore, our study is devoted to the IFE target fabrication with focusing on methods, which provide a high rep-rate and cost-effective target production. The top-level requirements necessary to achieve successful target fabrication and injection deal with target material selecting and fuel layer structuring. The following issues are of key importance:

1. For target mass-production those are:

    - Target materials must satisfy a wide range of required and desirable characteristics because the optimal microstructural design and material selection do allow one to obtain chemical, physical, and mechanical characteristics for specific applications.

    - Target fabrication capabilities and technologies must take into account the structure particularities of the solid structure of fusion fuel.

    - Fusion fuel must have an adequate thermal and mechanical stability for their quality survival in the process of target acceleration and delivery to the reaction chamber.

2. For cost-effective target production those are:

    - Minimization of time and space for all production steps.

    - Moving targets cooperate all production steps in the FST transmission line that is considered as a potential solution of passing from one-of-a-kind techniques to about 1 million targets per day.

    - Moving targets are the necessary condition for realizing a repeatable target production at required rates and their noncontact delivery at the laser focus.

3. For survivability of a fuel core (cryogenic layer) those are:

    - Layers with inherent survival features (fuel layer structuring—the grain size should be scaled back into the nanometer range).

    - Noncontact delivery system.

    - Multiple target protection methods including: (a) outer protective cryogenic layers, (b) metal coatings of different configurations and compositions, (c) nanocoatings for specific applications, and (d) coinjection of a special protective cover ahead of the target, etc.

For the IFE, all techniques must be integrated into an FST transmission line capable of producing about 1 million targets per day. Therefore, further studies are needed on MM-IFE connected to guide other R&D programs and to predict the behavior of IFE targets during their layering, delivery, and transport through the chamber environment. In addition, the MM-IFE allows reducing the cost of developments because it is intended to test the reactor-scaled technologies and to identify key issues for IFE commercialization. Implementation of MM-IFE program will be useful for working-out and substantiating the technical requirements needed for creation of a laser energy power plant.

Acknowledgements

This work was supported by the RF State Task of the Lebedev Physical Institute of Russian Academy of Sciences, and by the International Atomic Energy Agency (Research Contracts No. 20344, 11536, 13871).

Author details

Irina Aleksandrova[1], Elena Koresheva[1*], Evgeniy Koshelev[1], Boris Kuteev[2] and Andrei Nikitenko[1]

1 Lebedev Physical Institute of the RAS, Moscow, Russia

2 National Research Center "Kurchatov Institute", Moscow, Russia

*Address all correspondence to: elena.koresheva@gmail.com

IntechOpen

© 2018 The Author(s). Licensee IntechOpen. This chapter is distributed under the terms of the Creative Commons Attribution License (http://creativecommons.org/licenses/by/3.0), which permits unrestricted use, distribution, and reproduction in any medium, provided the original work is properly cited.

References

[1] Aleksandrova I, Koresheva E. Review on high rep-rate and mass-production of the cryogenic targets for laser IFE. High Power Laser Science and Engineering. 2017;**5**(2). e11(1-24)

[2] Fleurot N et al. The Laser Mégajoule (LMJ) project dedicated to inertial confinement fusion: Development and construction status. Fusion Engineering and Design. 2005;**74**(1):147-154

[3] Aleksandrova I et al. Mechanical mockup of IFE reactor intended for the development of cryogenic targets mass production and rep-rate delivery into the reaction chamber. In: 27th IAEA Fusion Energy Conference: Conference Proceedings, October 22–27, 2018, Mahatma Mondir Conference Center, Ahmedabad, India

[4] Koresheva E et al. HTSC maglev systems for IFE target transport applications. Journal of Russian Laser Research. 2014;**35**(2):151-168

[5] Aleksandrova I et al. Using HTSC-band of second generation for cryogenic transport of IFE targets. Bulletin of the Lebedev Physics Institute. 2015;**42**(11): 309312

[6] Aleksandrova I et al. A possibility to create the contact-free delivery system to transport the cryogenic hydrogen fuel for IFE power plants. Bulletin of the Lebedev Physics Institute. 2016;**43**(5): 160-166

[7] Aleksandrova I et al. Development of hybrid transport systems for delivering cryogenic fusion targets into focus of high-power laser system or ICF reactor. Journal of Russian Laser Research. 2017; **38**(3):249-264

[8] Aleksandrova I et al. Magnetic acceleration of the levitating sabot made of type-II superconductors. Journal of Russian Laser Research. 2018;**39**(2): 140-155

[9] Aleksandrova I et al. Progress in the development of tomographic information processing methods for applications to ICF target characterization. Fusion Technology. 2000;**38**(2):190-205

[10] Aleksandrova I et al. An efficient method of fuel ice formation in moving free standing ICF/IFE targets. Journal of Physics D: Applied Physics. 2004;**37**(8): 1163-1178

[11] Koresheva E et al. Possible approaches to fast quality control of IFE targets. Nuclear Fusion. 2006;**46**: 890-903

[12] Aleksandrova I et al. Cryogenic hydrogen fuel for controlled inertial confinement fusion (cryogenic target factory concept based on FST-layering method). Physics of Atomic Nuclei. 2017;**80**(7):1227-1248

[13] Tretyakov Y, Oleynikov N. Synthesis of functional nanocomposites based on solid-phase nanoreactors. Russian Chemical Reviews. 2004;**73**(9): 974-998

[14] Wanner R, Meyer H. Velocity of sound in solid hexagonal close-packed H_2 and D_2. Journal of Low Temperature Physics. 1973;**11**(5/6):715-744

[15] Aleksandrova I et al. Rapid fuel layering inside moving free-standing ICF targets: Physical model and simulation code development. Laser and Particle Beams. 2002;**20**:13-21

[16] Aleksandrova I et al. Survivability of fuel layers with a different structure under conditions of the environmental effects: Physical concept and modeling

results. Laser and Particle Beams. 2008; **26**(4):643-648

[17] Aleksandrova I et al. Thermal and mechanical responses of cryogenic targets with a different fuel layer anisotropy during delivery process. Journal of Russian Laser Research. 2008; **29**(5):419-431

[18] Aleksandrova IV et al. FST-technologies for high rep-rate production of HiPER scale cryogenic targets. Proceedings of SPIE. 2011;**8080**. 80802M (1-15)

[19] Aleksandrova I et al. Ultra-fine fuel layers for application to ICF/IFE targets. Fusion Science and Technology. 2013; **63**:106-119

[20] Aleksandrova I et al. Pilot target supply system based on the FST technologies: Main building blocks, layout algorithms and results of the testing experiments. Plasma and Fusion Research. 2013;**8**(2), 3404052:1-4

[21] Osipov I et al. A device for cryotarget rep-rate delivery in IFE target chamber. In: Tanaka K, Meyerhofer D, Meyer-ter-Vehn J, editors. Inertial Fusion Science and Applications, State of the Art 2001. Paris: Elsevier; 2002. pp. 810-814

[22] Aleksandrova I et al. Ultra-fine hydrogen layer fabrication in the conditions of out-vibrating influence on a cryogenic target. Voprosy Atomnoi Nauki i Tehniki. Seriya: Termoyadernyi Sintez; 2017;**40**(3):49-62 (in Russian)

[23] Sater J et al. Cryogenic DT-Fuel Layers Formed in 1 mm Spheres by Beta-Layering. Fusion Technology. 1999;**35**:229-233

[24] Bodner S et al. High Gain Target Design for Laser Fusion. Physics of Plasmas. 2000;7(6):2298-2301

[25] Landau L, Lifshitz E. Theoretical Physics. Electrodynamics of Continuous Media. Oxford, United Kingdom: Pergamon Press; 1960

[26] Lee S et al. Development and production of second generation high Tc superconducting tapes at SuperOx and first tests of model cables. Superconductors Science and Technology. 2014;**27**(4), 044022:1-9

[27] Urman Yu et al. On the levitation of diamagnetic bodies in the magnetic field. Technical Physics (Zhurnal Tekhnicheskoi Fiziki). 2010;**80**(9):25-33 (in Russian)

[28] Kislyak I et al. Investigations of superconductivity in MgB_2 bulk and Fe (steel)/MgB_2 wires. Voprosy Atomnoi Nauki I Tehniki. Series: Vacuum, Pure Materials, Superconductors. 2009;**18**: 107-110 (in Russian)

[29] Antonov Yu, Zaytsev A. Magneto Levitation Transport Technology. Moscow, Russian Federation: FIZMATLIT publishing house; 2014 (in Russian)

[30] Goldacker W et al. High transport currents in mechanically reinforced MgB_2wires. Superconductor Science and Technology. 2001;**14**(9):783-793

[31] Dolya S. Acceleration of magnetic dipoles by a sequence of current-carrying turns. Technical Physics. 2014; **59**(11):1694-1701

[32] Frank T, Pendergrass J. Power plant design for inertial confinement fusion: Implications for pellets. Journal of Vacuum Science and Technology. 1982; **20**(4):1381-1386

[33] Schlichting H. Boundary-Layer Theory. 7th ed. New-York, USA: McGraw-Hill, Inc.; 1979. ISBN: 0070553343

[34] Gophinath A, Koch D. Hydrodynamic interactions between two equal spheres in highly rarefied gas. Physics of Fluids. 1970;**11**:2772-2787

[35] Valmianski E et al. Wake shields for protection of IFE targets during injection. Fusion Science and Technology. 2003;**43**(3):334-338

Chapter 5

The Mechanical Behavior of the Cable-in-Conduit Conductor in the ITER Project

Donghua Yue, Xingyi Zhang and Youhe Zhou

Abstract

Cable-in-conduit conductor (CICC) has wide applications, and this structure is often served to undergo heat force-electromagnetic coupled field in practical utilization, especially in the magnetic confinement fusion (e.g., Tokamak). The mechanical behavior in CICC is of relevance to understanding the mechanical response and cannot be ignored for assessing the safety of these superconducting structures. In this chapter, several mechanical models were established to analyze the mechanical behavior of the CICC in Tokamak device, and the key mechanical problems such as the equivalent mechanical parameters of the superconducting cable, the untwisting behavior in the process of insertion, the buckling behavior of the superconducting wire under the action of the thermo-electromagnetic static load, and the Tcs (current sharing temperature) degradation under the thermo-electromagnetic cyclic loads are studied. Finally, we summarize the existing problems and the future research points on the basis of the previous research results, which will help the related researchers to figure out the mechanical behavior of CICC more easily.

Keywords: Nb3Sn, cable in conduit conductor (CICC), cable stiffness, coefficient of thermal expansion, untwisting, current sharing temperature

1. Introduction

The ITER program is one of the largest and the most influential international energy technology cooperation projects, to verify the engineering feasibility of the magnetic confinement fusion. The core device of the magnetic confinement fusion reactor is the cable-in-conduit conductor (CICC). CICCs were used to build up the superconducting coil for generating strong magnetic fields to confine the high-temperature plasma in a confined space and maintaining the fusion reaction [1]. The ITER superconducting magnet systems mainly consist of four kinds of coils: 6 central solenoids (CS), 18 toroidal field (TF) coils, 6 sets of poloidal field (PF) coils, and 9 pairs of correction fields coils (CC) [2].

As early as the 1960s, the low-temperature superconducting material NbZr was processed into round wire and cables [3]. Subsequently, the superconducting magnets were wound with the structure of an internally cooled conductor (ICS) [4]. The superconducting strand is cooled to the superconducting state by the heat transfer copper tube with liquid helium in it [5]. But, the contact cooling method of ICS is inefficient, and the superconducting material is inclined to have a magnetic

IntechOpen

flux jump which will make the magnetic system to be quenched. In 1975, Hoenig et al. suggested subdividing the superconductor into strands to suppress the flux jump and twisting them into a cable to reduce the AC losses [6]. In 1980, Lue et al. proposed a cable-in-conduit design, and the innermost part was a perforated copper tube or a high-hardness stainless steel spring to form a liquid helium fast-flowing channel [7]. These two designs are the prototypes of modern CICC conductors. Nowadays, the CS and TF conductors with higher magnetic fields in the ITER project were fabricated by more than 1000 Nb3Sn wires. Disadvantageously, the superconducting properties of Nb3Sn are sensitive to mechanical deformation, which means that the tensile, compressive, and torsional deformations all lead to the reduction of the critical current [8]. Therefore, the strain state of the Nb3Sn strand cannot be ignored. Therefore, during its design, manufacture, and operation stage of the CICC, the mechanical analysis is needed.

Many studies have been published on the equivalent mechanical parameters of the twisted cable with two dimensions, such as the compression modulus of the cross section [9, 10], rather than built a complete three-dimensional model of the twisted cable. Feng et al. have applied the thin rod model to CICC conductor analysis and established the spatial geometry of each superconducting strand in the CICC conductor [11]. Qin et al. have applied the thin rod model to the mechanical analysis of superconducting cables and derived the axial stress–strain curves of primary cables and high-order strands [12]. The influence of pitch on the elastic modulus of the stranded cable and the curvature of the strand has been discussed. The introduction of the copper strand has been found to greatly reduce the axial stiffness of the strand, and the contact deformation between the strands has been found to have a great influence on the stiffness of the strand. The theoretical calculation results have appeared to be in good agreement with the experiments [13, 14]. Yue et al. have conducted a systematic mechanical analysis of the CICC in the design, preparation, and operation stage [15–18].

In this chapter, first of all, we focus on the equivalent mechanical parameters of the superconducting cable; second, we concentrate on the untwisting behavior in the process of insertion; third, we want to explain the buckling behavior of the superconducting wire under the action of the thermo-electromagnetic static load; and finally, the Tcs degradation under the thermo-electromagnetic cyclic loads is studied. Our goal is to relate the cable stresses and buckling behavior to the thermal and electromagnetic loads so that relations between cable stress and current transport characteristics are built completely.

2. The equivalent mechanical parameters of the CICC

The mechanical behaviors of CICC have two main problems of structure and operation. On the one hand, the equivalent modulus of the cable is dependent on the manufacture parameters such as pitch, porosity, and radius. On the other hand, the electromagnetic load and the extremely low temperature make the internal stress and strain state of the cable difficult to analyze. Therefore, the thin rod model is applied to calculate the equivalent mechanical parameters of CICC conductors.

2.1 The tensile stiffness of the triplet

From the geometry characteristics of the cable, we know that the CICC superconducting cables have a complex structure with five stages of spirals.

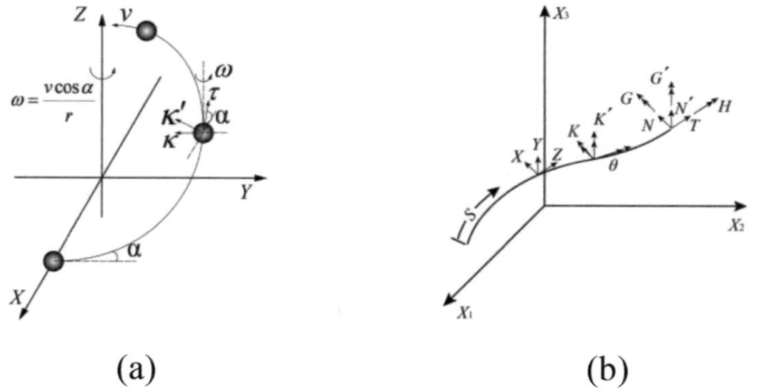

Figure 1.
The position of a spiral rod (a) and loads acting on the wire (b) [19].

Therefore, each wire in the cable can be simplified into a thin rod which is elongated in the axial direction under the axial tensile load, and the wires can be contacted tightly or rotate in the lateral direction. The deformation and the force analysis of the wires are shown in **Figure 1(a)** and **(b)**, respectively.

The curvature and torsion of the spiral are as follows:

$$\kappa = 0; \kappa' = \frac{v\mathrm{d}t\cos^2\alpha}{r\mathrm{d}s} = \frac{\cos^2\alpha}{r}; \tau = \frac{v\mathrm{d}t\cos\alpha\sin\alpha}{r\mathrm{d}s} = \frac{\cos\alpha\sin\alpha}{r}. \tag{1}$$

The equilibrium equation of forces and moments can be expressed as

$$\begin{aligned} &\frac{\mathrm{d}N}{\mathrm{d}s} + T\kappa' - N'\tau + X = 0; && \frac{\mathrm{d}G}{\mathrm{d}s} - G'\tau + Hk' - N' + K = 0; \\ &\frac{\mathrm{d}N'}{\mathrm{d}s} - Tk + N\tau + Y = 0; && \frac{\mathrm{d}G'}{\mathrm{d}s} - Hk + G\tau + N + K' = 0; \\ &\frac{\mathrm{d}T}{\mathrm{d}s} - Nk' + N'k + Z = 0; && \frac{\mathrm{d}H}{\mathrm{d}s} - Gk' + G'k + \Theta = 0. \end{aligned} \tag{2}$$

Assuming that the thin rod is isotropic and elastic, the moments in any cross section with respect to the axis x, y, z can be written as I_x, I_y, I_z, and the constitutive equations are given by

$$G = EI_x\Delta\kappa; G' = EI_y\Delta\kappa'; H = GI_z\Delta\tau. \tag{3}$$

In the result of solving Eq. (3) with the account for the temperature terms, the expression of the equivalent coefficient of thermal expansion in the axial direction of the triplet can also be given by [15]

$$\alpha_{\mathrm{eff}} = \frac{\alpha_{\mathrm{L}} - \alpha_{\mathrm{T}}\cos^2\alpha}{\sin^2\alpha}. \tag{4}$$

In Eq. (4), α_{eff} is the equivalent coefficient of thermal expansion of the strand, α_{L} is the coefficient of thermal expansion of the strand in the longitudinal direction, α_{T} is the transverse coefficient of thermal expansion of the strand, and α denotes the helix angle.

2.2 The tensile stiffness of the higher stage strand

Based on the equivalent modulus and thermal expansion of the triplet, the space and the 2D view of the triplet and single wire are shown in **Figure 2(a)** and **(b)**, respectively.

The conversion relationship between the local coordinates of the triplet and the higher-level strand can be expressed as

$$T_{\mathrm{k}} = \begin{bmatrix} -\cos\theta_{\mathrm{k}} & -\sin\theta_{\mathrm{k}} & 0 \\ \sin\theta_{\mathrm{k}}\sin\alpha_{\mathrm{k}} & -\cos\theta_{\mathrm{k}}\sin\alpha_{\mathrm{k}} & \cos\alpha_{\mathrm{k}} \\ -\sin\theta_{\mathrm{k}}\cos\alpha_{\mathrm{k}} & \cos\theta_{\mathrm{k}}\cos\alpha_{\mathrm{k}} & \sin\alpha_{\mathrm{k}} \end{bmatrix}. \tag{5}$$

The curvature and torsion of the secondary stage strand can be given by

$$\begin{bmatrix} \kappa_{\mathrm{p2}} \\ \kappa_{\mathrm{b2}} \\ \kappa_{\mathrm{t2}} \end{bmatrix} = T_2 T_1 \left\{ T_1^{\mathrm{T}} \begin{bmatrix} 0 \\ 0 \\ \dfrac{\cos\alpha_2}{r_2} \end{bmatrix} + \begin{bmatrix} 0 \\ 0 \\ \dfrac{\cos\alpha_1}{r_1} \end{bmatrix} \sin\alpha_2 \right\}. \tag{6}$$

According to the geometric compatibility of the secondary-stage strand, the deformation of the triplet is equal to the tangential strain of the secondary-stage strand, and the torsion of the triplet is equal to the twist angle of the secondary cable. The axial loads and torque of the secondary-stage strand can be obtained. The equilibrium equations can be expressed as

$$\begin{bmatrix} 0 \\ 0 \\ F_{\mathrm{t0}} \end{bmatrix} = 3^{*}3^{*} T_1 T_2 \begin{bmatrix} F_{\mathrm{p2}} \\ F_{\mathrm{b2}} \\ F_{\mathrm{t2}} \end{bmatrix} \tag{7}$$

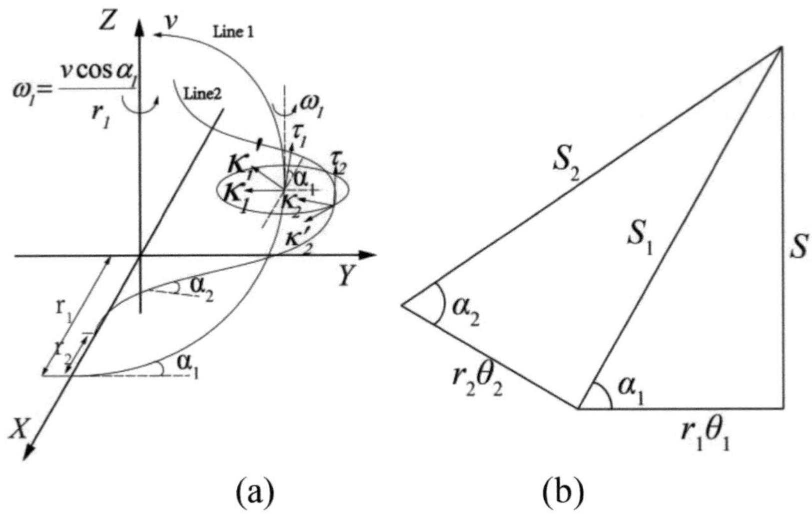

Figure 2.
Space line of the triplet and single wire (a) and 2D view of the triplet and single wire (b).

$$\begin{bmatrix} M_{p1} \\ M_{b1} \\ M_{t1} \end{bmatrix} = 3T_2 \begin{bmatrix} M_{p2} \\ M_{b2} \\ M_{t2} \end{bmatrix} + 3 \begin{bmatrix} r_2(F_{t2}\sin\alpha_2 + F_{b2}\cos\alpha_2)\sin\theta_2 \\ -r_2(F_{t2}\sin\alpha_2 + F_{b2}\cos\alpha_2)\cos\theta_2 \\ r_2(F_{t2}\cos\alpha_2 - F_{b2}\sin\alpha_2) \end{bmatrix}. \tag{8}$$

Theoretically, the tensile stiffness of the conductor can be deduced by four times transformations as the CICC conductor has a five-stage twist structure.

3. Rotation analysis of the CICC

In the CICC conductor manufacture process, they twist a superconducting cable and penetrate it into the stainless steel tube. However, due to the friction between the superconducting cable and the stainless steel armor, the drag force of the cable is as high as several tons during the cable penetration. The friction force of the pipe leads to the axial elongation of the superconducting cable, accompanied by the untwisting of the cable, which causes the cable pitch to increase. This makes that the pitch is much larger than the ITER requirement [20]. Therefore, the untwist behavior of the cable must be controlled [17].

In this section, the untwist model is described. The large-scale cable is considered, e.g., ITER TF, CS, and CFETR CSMC. The components of the final cable include petals, central cooling spiral, and wrap, as shown in **Figure 3**. The model ignores the friction between the jacket and the cable, only modeling de-twists of the cable under the insertion force F_{Insert}.

The cable is divided into three parts in the model: central cooling spiral, six petals, and the wrap. The twist direction of wrap and cooling spiral is left and with the reverse direction for the petal. The torsion constraint is free for the cable when there is undering the uniaxial tension. Therefore, the boundary conditions can be set as $F = F_0; M = 0$. The force of the whole cable is from those acts on wrap, sub-cables, and central cooling spiral, which can be described as follows:

$$\begin{aligned} F &= F_{in} + F_p + F_{st} = F_0, \\ M &= M_{in} + M_p + M_{st} = 0. \end{aligned} \tag{9}$$

In Eq. (9), F is the insertion force in the axial direction for the cable. F_{st}, F_p, F_{in} are the forces loading on the stainless steel wrap, petals, and inner cooling spiral,

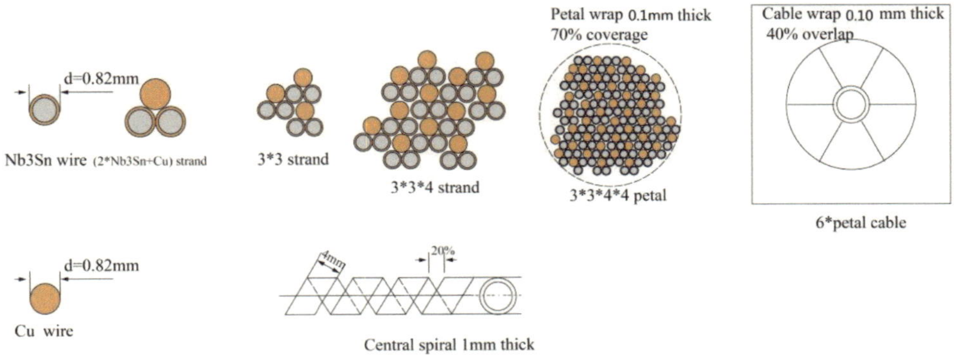

Figure 3.
The dimensions and parameters of a large-scale cable (e.g., CFETR CSMC).

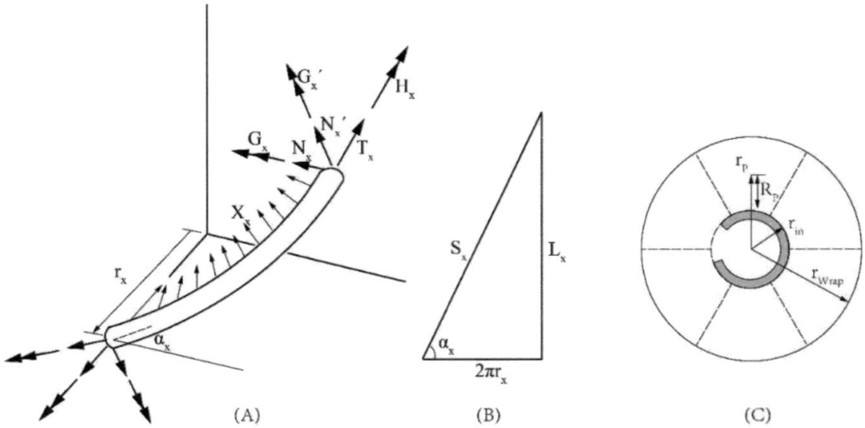

Figure 4.
Force distribution (A) and spatial relationship (B and C).

respectively; M represents the torque of the whole cable. $M_{\text{st}}, M_p, M_{\text{in}}$ denote the torques of stainless steel wrap, petals, and inner cooling spiral.

The force distribution and spatial relationship are shown in **Figure 4**. G_x, G_x', H_x are sectional moment components of the thin rod. N_x, N_x', T_x are the sectional force components. X_x is the contact force. α_x is the spiral angle of each component. r_x is the distance between the centroid and the center of the cable, and L_x is the twist pitch.

We simplified the central cooling spiral into a single helical thin plate. The change of spiral angle, radius, and the axial strain are defined as $\Delta\alpha_{st}, \frac{\Delta r_{st}}{r_{st}}, \xi_{st}$, respectively. Then, the geometrical equations of the He-inlet spiral, six petals, and stainless steel wrap can be deduced by Costello-Velinsky theory [19]; we can obtain

$$He-inlet: \begin{cases} \varepsilon_0 = \xi_{in} + \dfrac{\Delta\alpha_{in}}{\tan\alpha_{in}}, \\ \beta_0 = \dfrac{\xi_{in}}{\tan\alpha_{in}} + \dfrac{1}{\tan\alpha_{in}}\dfrac{\Delta r_{in}}{r_{in}} - \Delta\alpha_{in}, \end{cases}$$

$$Petal: \begin{cases} \varepsilon_0 = \xi_P + \dfrac{\Delta\alpha_P}{\tan\alpha_P}, \\ \beta_0 = \dfrac{\xi_P}{\tan\alpha_P} + \dfrac{1}{\tan\alpha_P}\dfrac{\Delta r_P}{r_P} - \Delta\alpha_P, \end{cases} \tag{10}$$

$$Wrap-stiffness: \begin{cases} \varepsilon_0 = \xi_{st} + \dfrac{\Delta\alpha_{st}}{\tan\alpha_{st}}, \\ \beta_0 = \dfrac{\xi_{st}}{\tan\alpha_{st}} + \dfrac{1}{\tan\alpha_{st}}\dfrac{\Delta r_{st}}{r_{st}} - \Delta\alpha_{st}. \end{cases}$$

In the equations, the axial deformation and torsion angle of the cable are expressed as ε_0, β_0. The only axial tension is considered for central cooling spiral and wrap. Therefore, the equations can be updated as

$$
\begin{aligned}
&He-inlet:\begin{cases} G_{in}=0; G'_{in}=0; H_{in}=0 \\ T_{in}=E_{in}A_{in}\xi_{in} \end{cases} \\
&Petal:\begin{cases} G_P=0; G'_P=E_PI^x_P\Delta\kappa'_P; H_P=G_PI^z_P\Delta\tau_P \\ T_P=E_PA_P\xi_P \end{cases} \\
&Wrap-stiffness:\begin{cases} G_{st}=0; G'_{st}=0; H_{st}=0 \\ T_{st}=E_{st}A_{st}\xi_{st} \end{cases}
\end{aligned}
\tag{11}
$$

In the equation systems, E refers to Young's modulus of each component. A refers to the cross section of each component. $E_PI^x_P$, G_PI_P are the bending and torsion stiffness of petals, respectively. $\Delta\kappa'_P$, $\Delta\tau_P$ are the changes in curvature and twist, respectively. ξ_P is the axial strain of petal.

The balance equations and the compatible equations of central cooling spiral, subcables, and wrap can be obtained as

$$
\begin{aligned}
N'_x &= G'_x\tau_x + H_x\kappa'_x, \\
X_x &= N'_x\tau_x - T_x\kappa'_x, \\
F_x &= T_x\sin\alpha_x + N'_x\cos\alpha_x, \\
M_x &= H_x\sin\alpha_x + G'_x\cos\alpha_x + T_xr_x\cos\alpha_x - N'_xr_x\sin\alpha_x.
\end{aligned}
\tag{12}
$$

$$
\begin{aligned}
r_{in} &= R_{in}; r_P = R_{in} + R_P; r_{st} = R_{in} + 2R_P, \\
\Delta r_{in} &= \Delta R_{in}; \Delta r_P = \Delta R_{in} + \Delta R_P; \Delta r_{st} = \Delta R_{in} + 2\Delta R_P.
\end{aligned}
\tag{13}
$$

where x stands for anyone of in, p, st, which represent inner He-inlet spiral, petal, and stainless steel wrap, respectively; R_p denote the twist radius of petal, and $\Delta R_P = -\nu R_P\xi_P$, v is Poisson's ratio of the petal.

The contact force loading on the petal can be written as follows:

$$
\frac{6X_P}{\sin\alpha_P} = \frac{X_{st}}{\sin\alpha_{st}} - \frac{X_{in}}{\sin\alpha_{in}}
\tag{14}
$$

where X_P stand for the line pressure between the petal and inner He-inlet spiral; X_{st} is the uniform force between the stainless steel wrap and petal; and X_{in} is the reaction force between inner He-inlet spiral and petal.

From Eqs. (9)–(12), one can get

$$
\begin{aligned}
F &= F_0 = 6\left(T_P\sin\alpha_P + N'_P\cos\alpha_P\right) + T_{st}\sin\alpha_{st} + T_{in}\sin\alpha_{in} \\
M &= 0 = 6\left(H_P\sin\alpha_P + G'_P\cos\alpha_P + T_Pr_P\cos\alpha_P - N'_Pr_P\sin\alpha_P\right) \\
&\quad -T_{st}r_{st}\cos\alpha_{st} - T_{in}r_{in}\cos\alpha_{in}
\end{aligned}
\tag{15}
$$

From the above Eqs. (10)–(15), we can obtain

$$
\begin{aligned}
&\left(\frac{1}{\tan\alpha_{st}}+\tan\alpha_p\right)\xi_{st}-\frac{\upsilon r_p}{\tan\alpha_{st}}\frac{2}{r_{in}+2R_P}\xi_P+0\cdot\xi_{in}+\frac{1}{\tan\alpha_{st}}\frac{1}{r_{in}+2R_P}\Delta r_{in}-\tan\alpha_{st}\varepsilon_0-\beta_0=0\\
&0\cdot\xi_{st}+\left(\frac{1}{\tan\alpha_p}+\tan\alpha_p-\frac{\upsilon R_P}{\tan\alpha_P}\frac{1}{r_{in}+R_P}\right)\xi_P+0\cdot\xi_{in}+\frac{1}{\tan\alpha_P}\frac{1}{r_{in}+R_P}\Delta r_{in}-\tan\alpha_P\varepsilon_0-\beta_0=0\\
&0\cdot\xi_{st}+0\cdot\xi_P+\left(\tan\alpha_{in}+\frac{1}{\tan\alpha_{in}}\right)\xi_{in}+\frac{1}{r_{in}\tan\alpha_{in}}\Delta r_{in}-\tan\alpha_{in}\varepsilon_0-\beta_0=0\\
&-\frac{\sin\alpha_P}{6\sin\alpha_{st}}(EA)_{st}\kappa'_{st}\xi_{st}+\left[GB\tau_P+HB\kappa'_P-(EA)_P\kappa'_P\right]\xi_P+\frac{\sin\alpha_P}{6\sin\alpha_{in}}(EA)_{in}\kappa'_{in}\xi_{in}+\left(GD\tau_P+HD\kappa'_P\right)\Delta r_{in}\\
&+\left(GE\tau_P+HE\kappa'_P\right)\varepsilon_0+0\cdot\beta_0=0\\
&(EA)_{st}\sin\alpha_{st}\xi_{st}+\left[6GB\tau_P\cos\alpha_P+6HB\kappa'_P\cos\alpha_P+6(EA)_P\sin\alpha_P\right]\xi_P+(EA)_{in}\sin\alpha_{in}\xi_{in}\\
&+6\left(GD\tau_P+HD\kappa'_P\right)\cos\alpha_P\Delta r_{in}+6\left(GE\tau_p+HE\kappa'_P\right)\cos\alpha_P\varepsilon_0+0\cdot\beta_0=F_0\\
&-(EA)_{st}\cos\alpha_{st}r_{st}\xi_{st}+\left[6HB\sin\alpha_P+6GB\cos\alpha_P+6(EA)_P(r_{in}+r_P)\cos\alpha_P-6\left(GB\tau_P+HB\kappa'_P\right)r_P\sin\alpha_P\right]\xi_P\\
&-(EA)_{in}\cos\alpha_{in}r_{in}\xi_{in}+\left[6HD\sin\alpha_P+6GD\cos\alpha_P-6\left(GD\tau_P+HD\kappa'_P\right)r_P\sin\alpha_P\right]\Delta r_{in}\\
&+\left[6HE\sin\alpha_P+6GE\cos\alpha_P-6\left(GE\tau_P+HE\kappa'_P\right)r_P\sin\alpha_P\right]\varepsilon_0+0\cdot\beta_0=M_0
\end{aligned}
\tag{16}
$$

The twist angle β_0 can be computed from the equation system (16). So, the rotation of cable can be evaluated [17].

The experimental and numerical results are shown in **Figure 5**. First, it is easily found that the cabling tension has less impact on cable rotation. Taking the wrapping tension with 200 N, for example, there is no deviation between two different cables with cabling tension 200 N and 800 N, respectively. This result is in good agreement with the numerical model results. Second, the untwisting of the cable was mainly controlled by wrapping tension. Therefore, we can reduce the rotation significantly by increasing the wrapping tension. When insertion force is about 40 kN, the untwisting angle of cables with 600 N wrapping tension is about half of those cases with 200 N.

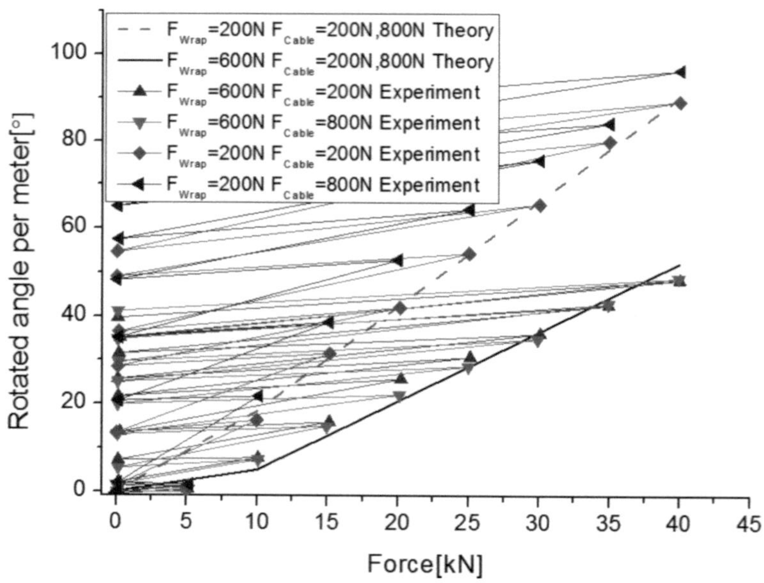

Figure 5.
Rotated angle per meter as a function of the force: Numerical and experimental results.

4. Buckling behavior in the CICC

4.1 Analytical model without the electromagnetic force

It is known that all the ITER CS and TF coils are wounded by CICCs, which made up of five-stage sub-cables formed around a central cooling tube. The petals and the sub-cables are wrapped with stainless steel tapes. Then, let the wrapped cable inserted into a stainless steel tube, which act as an amor. We assume that the total length of free segment of the superconducting strand on the surface of the cable is L (equal to the twist pitch), and set the fixed constraints on both sides, the wrap band as a uniform spring constraint. The schematic of this analytical model is illustrated in **Figure 6**. The lengths of spring constraint on both sides are equal to L_1 and L_3, respectively. The length of the free fraction of the strand is of L_2; we can get $L_1 + L_2 + L_3 = L$.

Since the coefficient of thermal expansion of stainless steel between 923 and 4.2 K is approximately twice that of the Nb3Sn strand, then the superconducting cable is in compression at the end of the cooling. The thermal shrinkage of the cable is denoted by $\varepsilon_{Thermal}$. Other than this, the strands of the cable can be squeezed into another side of the conduit by the large electromagnetic force; then, there will generate a large void on the other side of the conduit. Due to the gap, the friction force between the strands and the stainless steel armor decrease significantly. As there is no lateral restraint by wrap or friction, the surface strands around the void will show bending deformation by the thermal mismatch. In addition, the strand will slide into the high magnetic field region that will aggravate its bending behavior. ε_{Slid} is the stand for this slid strain. Therefore, the total compression strain of the strand ε_T is the sum of ε_{Slid} and $\varepsilon_{Thermal}$. In this case, $\varepsilon_T = \varepsilon_{Thermal} + \varepsilon_{Slid}$.

The mechanical analysis of the strand microelement is shown in **Figure 7**. The equilibrium equations for the moments are as follows:

$$\frac{dM}{dx} + N\frac{dy}{dx} - Q_v = 0 \tag{17}$$

In Eq. (17), Q_v, M, N denote the vertical shear force, the bending moment, and the compression force along the axial direction, respectively. After submitting $M = -EI\left(d^2y/dx^2\right)$ into Eq. (17), and making a substitution $dQ_v/dx = ky$, we can get the differential equation of the rod with the spring constraints:

$$EI\frac{d^4y}{dx^4} - N\frac{d^2y}{dx^2} + ky = 0 \tag{18}$$

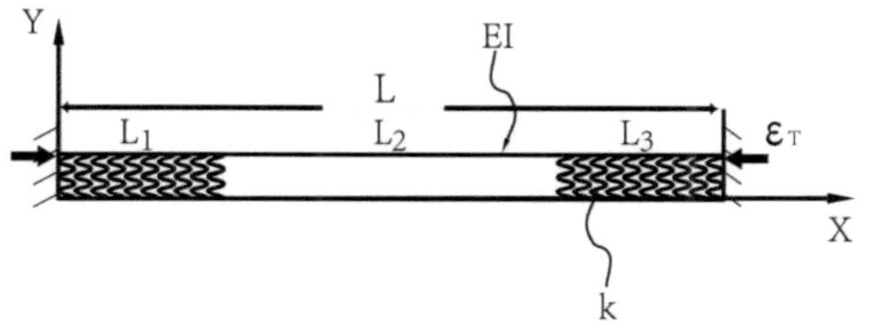

Figure 6.
Schematic of the strand model ignores the EM force. L, L_1, L_2 and L_3 denote the length of the twist pitch, left spring constraint, free segment, and right spring constraint, respectively. k denotes the rigidity of the bandaging. EI is the bending stiffness of the strand. ε_T is the total compression strain, which is equal to the sum of $\varepsilon_{Thermal}$ and ε_{Slid}.

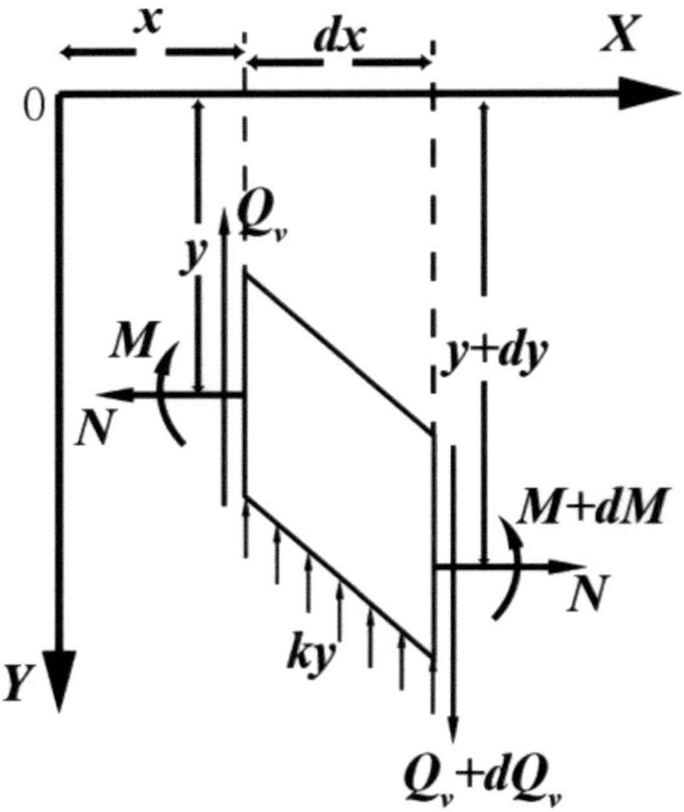

Figure 7.
Mechanical analysis for the strand microelement.

If the strand has no wrapping, one can get

$$EI\frac{d^4y}{dx^4} - N\frac{d^2y}{dx^2} = 0 \tag{19}$$

where k denotes the stiffness of the wrapping.

The continuity of displacements, twist angles, bending moments, and shear forces across the coverage/free strand require

$$\begin{aligned} w_i &= w_{i+1} \\ \frac{dw_i}{dx_i} &= \frac{dw_{i+1}}{dx_{i+1}} \\ \frac{d^2w_i}{dx_i^2} &= \frac{d^2w_{i+1}}{dx_{i+1}^2} \\ -\left[N\frac{dw_i}{dx_i} + \frac{d^3w_i}{dx_{i+1}^3}\right] &= -\left[N\frac{dw_{i+1}}{dx_{i+1}} + \frac{d^3w_{i+1}}{dx_{i+1}^3}\right] \end{aligned} \tag{20}$$

In Eq. (20) the subscript i represents the number of wrapping/free segments. The fixed boundary conditions at the coverage fraction have the form:

$$w|_{x=0} = w|_{x=L} = 0$$
$$\left.\frac{dw}{dx}\right|_{x=0} = \left.\frac{dw}{dx}\right|_{x=L} = 0 \tag{21}$$

Thus, the general solution of Eq. (18) is obtained as

$$w_i = a_{i1}\cos(\lambda x_i) + a_{i2}\sin(\lambda x_i) + a_{i3}x_i + a_{i4} \tag{22}$$

This general solution can be divided into three situations with a variation of the external force P [21–23]:

1. If $P \le 2\sqrt{EIk}$, the general solution has the form as

$$w_i = a_{i1}\cosh(\lambda_1 x_i) + a_{i2}\cosh(\lambda_2 x_i) + a_{i3}\sinh(\lambda_1 x_i) + a_{i4}\sinh(\lambda_2 x_i) \tag{23}$$

2. If $P = 2\sqrt{EIk}$, it becomes

$$w_i = a_{i1}\cos(\lambda x_i) + a_{i2}\cos(\lambda x_i) + a_{i3}x_i\sin(\lambda x_i) + a_{i4}x_i\sin(\lambda x_i) \tag{24}$$

3. If $P \ge 2\sqrt{EIk}$, it becomes

$$w_i = a_{i1}\cos(\lambda_1 x_i) + a_{i2}\cos(\lambda_2 x_i) + a_{i3}\sin(\lambda_1 x_i) + a_{i4}\sin(\lambda_2 x_i) \tag{25}$$

Here, λ, λ_1, λ_2 and λ_3 are the eigenvalues, respectively; and a_{ij} is the constant coefficients.

The axial compression strain can be neglected when the strand gets into the buckling state, as the compression force N is small. Therefore, the total compression of the strand can be expressed as

$$\frac{1}{2}\int_0^L (w')^2 dx = L\varepsilon_T - \frac{NL}{EA} \tag{26}$$

Here, w' denotes the curvature of the strand transverse deflection. The radius of curvature ρ has the form

$$\rho = \frac{d^2w}{d^2x} \tag{27}$$

Here, w stands for the strand transverse deflection. The corresponding maximum strain is

$$\varepsilon_{\max} = \frac{D}{2\rho} \tag{28}$$

Here, D stands for the diameter of the strand. If the maximum strain is larger than 1%, the strand would be considered as cracking [10, 24–26]. Based on these equations, the coefficient a_{ij} can be calculated, as well as the buckling deflection, the relationship between the radius of curvature, and the thermal compression strain $\varepsilon_{Thermal}$ or slid strainε_{Slid}.

4.2 Analytical model with the electromagnetic force

As we know, the magnetic field is maintained at 13 T in the Tcs test procedure, and the electromagnetic force rises with the increasing the current. Therefore, the electromagnetic force is a cyclic load. The strand on the surface of the cable where the gap is formed is selected in this model. The EM force F_{EM} is perpendicular to the Nb3Sn strand, and its direction points to the inner part of the cable. Thus, the strand at the lower loading side has the least lateral constraint that means the strand most likely to have a buckling behavior in the lateral direction. Therefore, the only thing that can prevent the buckling of the strands is the friction force against the cable. The friction factor can be written as a symbol μ.

In **Figure 8**, μ_L and μ_T are the axial friction factor and lateral friction factor, respectively; L_s and L denote the slipping length and buckling length, respectively; $L + 2L_s$ is the twist pitch of the first stage; F_{EM} stands for the EM force, and then in the buckling area $\mu_T F_{EM}$ is the lateral constraint on the strand; and P_0 and P are the compression force of the strand in the slipping and buckling area, respectively.

Accordingly, by the torque balance of the microelement, as shown in **Figure 9**. The equilibrium equation is as follows:

$$-EIy'' = Py + \frac{\mu_T F_{EM}}{2}\left(\frac{L}{2} - x\right)\left(\frac{L}{2} - x\right) - \frac{\mu_T F_{EM} L}{2}\left(\frac{L}{2} - x\right) \tag{29}$$

In Eq. (29) EI stands for the bending stiffness of the Nb_3Sn strand; y stands for the deflection of the buckling. After simplifying formula (29), one can get

$$EIy'' + Py + \frac{\mu_T F_{EM}}{2}\left(\frac{L^2}{4} - x^2\right) = 0 \tag{30}$$

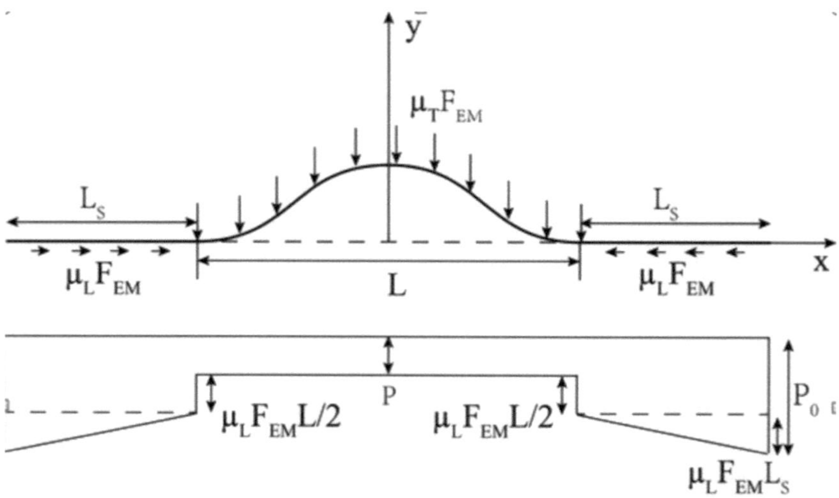

Figure 8.
Schematic of the analytical model with the EM force.

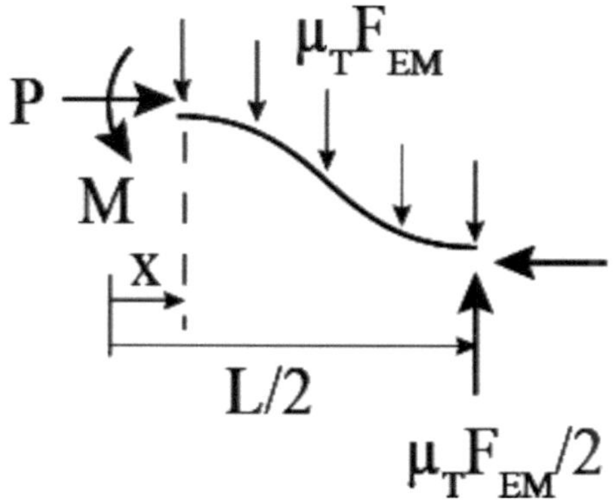

Figure 9.
Schematic view of the mechanical analysis for the microelement.

The general solution [27] of formula (30) is given as

$$y = \frac{m}{n^4}\left[A\cos(nx) + B\sin(nx) - \frac{n^2x^2}{2} + \frac{n^2L^2}{8} + 1\right] \tag{31}$$

In Eq. (31), $m = \frac{\mu_T F_{EM}}{EI}$, $n^2 = \frac{P}{EI}$. Two boundary conditions are needed to solve the undetermined coefficients. After applying $y|_{x=\pm L/2} = 0, y'|_{x=0} = 0$ to formula (31), one can get

$$y = \frac{m}{n^4}\left[\frac{-\cos(nx)}{\cos(nL/2)} - \frac{n^2x^2}{2} + \frac{n^2L^2}{8} + 1\right] \tag{32}$$

If we substitute the boundary $y'|_{x=\pm L/2} = 0$ to Eq. (32), Eq. (32) becomes

$$y' = \tan(nL/2) - nL/2 = 0 \tag{33}$$

We can deduce that $nL = 8.9868...$ from Eq. (33). Therefore, the axial compression force in the buckling area is $P = 80.76EI/L^2$. There must be a balance in the axial direction between the buckling and slipping segments; we can get that

$$P_0 - P = \frac{\mu_L F_{EM} L}{2} + \mu_L F_{EM} L_S \tag{34}$$

The geometric compatibility equation of the strand can be expressed as

$$\frac{(P_0 - P)L}{EA} - \int_{-L/2}^{L/2} \frac{1}{2} y'^2 dx + \frac{\mu_L F_{EM} L_S^2}{EA} = 0 \tag{35}$$

Submitting Eq. (32) into Eq. (35), one can get the relation between axial force and the length of the buckling area:

$$\frac{(P_0 - P)L}{EA} - 1.597 \times 10^{-5}\left(\frac{\mu_T F_{EM}}{EI}\right)^2 L^7 + \frac{\mu_L F_{EM} L_S^2}{EA} = 0 \tag{36}$$

Submitting Eq. (34) into Eq. (36) and eliminating the L_s, one can get $P_0 = P + \sqrt{1.597 \times 10^{-5} EA\mu_L F_{EM} \left(\frac{\mu_T F_{EM}}{EI}\right)^2 L^7 - \frac{L^2}{4}(\mu_L F_{EM})^2}$, in which $P = 80.76EI/L^2$.

Consider the special case $L_S = 0$ that means the end of the strand is locked and the length of the buckling is equal to the twist pitch. In this case Eq. (36) can be simplified as follows:

$$P_0 = 80.76\frac{EI}{L^2} + 1.597 \times 10^{-5} EA \left(\frac{\mu_T F_{EM}}{EI}\right)^2 L^6 \tag{37}$$

Based on these equations, the critical buckling load P_0 can be calculated, as well as the relationship between the buckling length, bending stiffness, and the friction factor.

We can know that the strand buckling behavior is depending on the twist pitch of the first stage; the shorter the lay length, the lower the possibility of the strand buckling. The higher the wrap rigidity, the stronger the strand. Furthermore, if we fixed the coverage rate, and with a narrow wrap, which would lead to almost uniform lateral supporting, it can also prevent the strand slid into buckling. This mechanism of buckling is shown in **Figure 10**.

When there is no thermal load and EM force, the original strand in the CICCs is shown in **Figure 10(a)**. When the temperature is cooled down, the axial strain caused by thermal mismatch and the lateral compression raised by the EM load are applied to the strands. We can see that the initial "straight" strand was keeping its shape by well lateral support. When the strand working in a large magnetic field (including self-field), it bears a huge EM force. A large void is generated, at the same time, some strands bending, which is shown in **Figure 10(b)**. Additionally, this bending strain is not the only factor to make the strand fracture. For the SULTAN measurements [28], the samples is about 3.6 m long, and the high-field

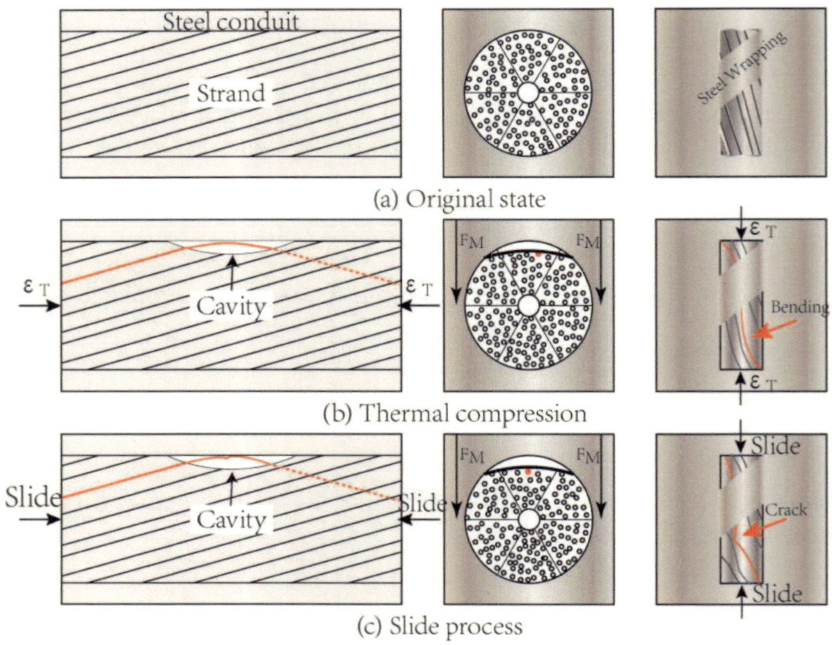

Figure 10.
Mechanism of the strand buckling behavior during the cooling and conducting processes.

region is about 0.4 m. During the EM cyclic loading, the cable can slide into the high-field region, which can accelerate the wire bending and leading the strand to fracture. This process is shown in **Figure 10(c)**.

5. Mechanical behavior of the CICC caused by electromagnetic force and thermal mismatch

The CICC qualification test samples show gradual degradation of the current-sharing temperature (Tcs) under several hundreds of EM cycles [29, 30], which leads to the Nb_3Sn strand's bending or compressing deformation. In this section, we focus on the relationship between Tcs and axial strain of the cable.

It is known that the ITER CS and TF coils are wounded by CICCs, which made up of five-stage sub-cables formed around a central cooling tube. The fourth petals and the sub-cables are wrapped with stainless steel tapes. Then, the wrapped cable were inserted into a stainless steel tube, which act as an amor, as illustrated in the cross-sectional view in **Figure 11**. The inner diameter of the CS armor is equal to 36 and 37 mm for TF, respectively. The side length of CS conductor is 49 mm, and the diameter of the TF conductor is of 40 mm. More than 1000 Nb3Sn strands were in the CICC conductors. In this section, the conductors can be simplified into a rope and the petals with circle cross section, as shown in **Figure 11**. Thus, the cable could be analyzed by using the thin rod model as shown in **Figure 12**.

During the calculations, the parameters R, R_{in}, and twist pitch (h) are 6, 6, and 450 mm, respectively. According to the geometric relation $h = 2\pi r \tan\alpha$, as shown in **Figure 13(b)**, we can get $\alpha = 80.5°$, where α represents the initial helix angle. **Figure 13(a)** also displays the loads acting on the petal and the geometric relation of the petal centerline.

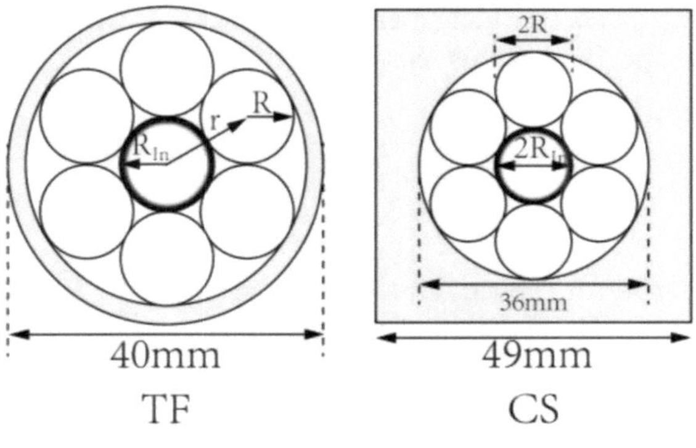

Figure 11.
Schematic illustration of the CS and TF CICC cross sections. The symbols R_{in} and R denote the radius of the He channel and petal, respectively; r is on behalf of the sum of R_{in} and R: $r = R_{In} + R$.

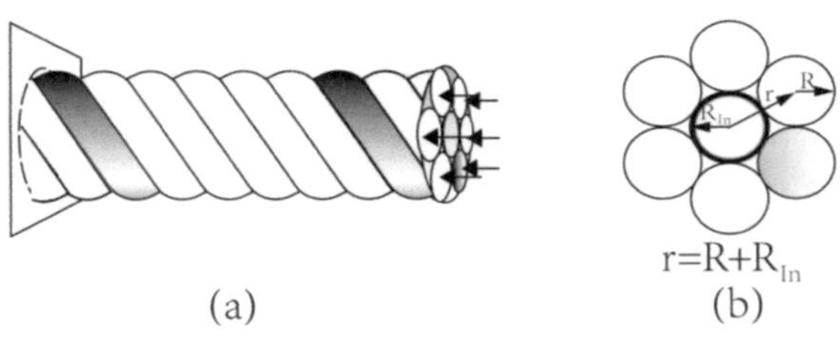

Figure 12.
(a) The deformation of the cable caused by the axial compression; (b) the cross section of the cable.

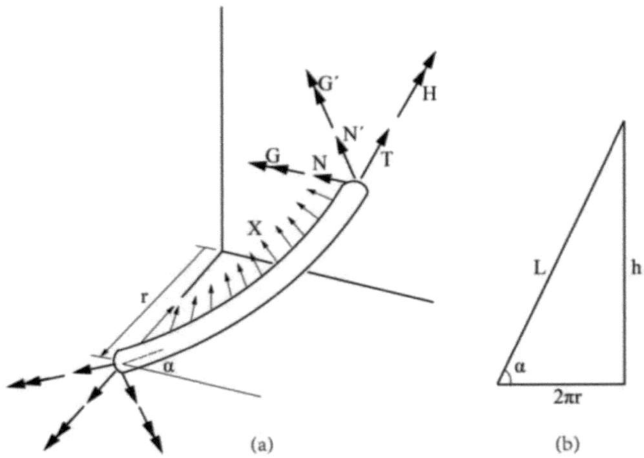

Figure 13.
(a) Uniform forces acting on the petal; (b) geometric relation of the petal centerline.

Assume that the petals were contact with each other in the original state. The curvature and the twist per unit length of the petal is k, k', τ; then the changes can be written as Δk, $\Delta k'$, $\Delta\tau$. They can be expressed as [19]

$$\Delta\kappa' = \frac{\cos^2\overline{\alpha}}{\overline{r}} - \frac{\cos^2\alpha}{r} = \frac{-\cos^2\alpha}{r}\frac{\Delta r}{r} - 2\frac{\sin\alpha\cos\alpha}{r}\Delta\alpha \tag{38}$$

$$\Delta\tau = \frac{\cos\overline{\alpha}\sin\overline{\alpha}}{\overline{r}} - \frac{\cos\alpha\sin\alpha}{r} = \frac{-\sin\alpha\cos\alpha}{r}\frac{\Delta r}{r} + \frac{(1-2\sin^2\alpha)}{r}\Delta\alpha \tag{39}$$

In Eqs. (38) and (39), α and $\overline{\alpha}$ is the initial and final helical angle of the petal, $\Delta\alpha$ and Δr stand for the change of helical angle and radius of the petal, and r and $\overline{r}$ denote the original and final radius, respectively. The loads and moments can be deduced as.

$$G' = \frac{E\pi R^4}{4}\Delta\kappa'; H = \frac{E\pi R^4}{4(1+\nu)}\Delta\tau; T = \pi E R^2\xi \tag{40}$$

Here, T, G', H stand for the axial load, bending moment, and twist moment of the petal, respectively; E is of the axial stiffness of the petal, v is Poisson's ratio, and ξ stands for the strain in the axial direction of the petal. If the petal is free in the initial state and there is no contact force under the compression state, then the uniform load between the petals is equal to zero. According to the thin rod model presented by Costello [19], the following equation is satisfied:

$$X = N'\frac{\sin\alpha\cos\alpha}{r} - T\frac{\cos^2\alpha}{r} = 0 \tag{41}$$

$$N' = H\frac{\cos^2\alpha}{r} - G'\frac{\sin\alpha\cos\alpha}{r} \tag{42}$$

In Eqs. (41) and (42), X stands for the resultant contact force per unit length of a petal; N' is of the shear force acting on the petal. From Eqs. (38)–(40) and Eq. (42), the shearing force N' acting on the petal can be written as

$$\frac{N'}{ER^2} = \frac{\pi v\sin\alpha\cos^3\alpha}{4(r/R)^2(1+v)}\frac{\Delta r}{r} + \frac{\pi\cos^2\alpha}{4(r/R)^2}\left(\frac{1-2\sin^2\alpha}{1+v} + 2\sin^2\alpha\right)\Delta\alpha \tag{43}$$

Substituting Eqs. (40) and (43) into Eq. (41), the axial strain of the petal ξ which is only related to the Δr and $\Delta\alpha$ can be written as

$$\xi = \frac{\upsilon \sin^2\alpha \cos^2\alpha}{4(r/R)^2(1+\upsilon)}\frac{\Delta r}{r} + \frac{\sin\alpha\cos\alpha}{4(r/R)^2}\left(\frac{1-2\sin^2\alpha}{1+\upsilon} + 2\sin^2\alpha\right)\Delta\alpha \quad (44)$$

The deformed configuration of the petal in **Figure 13** yields

$$\varepsilon = \frac{\overline{h} - h}{h} = \xi + \frac{\Delta\alpha}{\tan\alpha} \quad (45)$$

From Eqs. (44) and (45), one can get

$$\varepsilon = \frac{\upsilon \sin^2\alpha \cos^2\alpha}{4(r/R)^2(1+\upsilon)}\frac{\Delta r}{r} + \left[\frac{\cos\alpha}{\sin\alpha} + \frac{\sin\alpha\cos\alpha}{4(r/R)^2}\left(\frac{1-2\sin^2\alpha}{1+\upsilon} + 2\sin^2\alpha\right)\right]\Delta\alpha \quad (46)$$

The angle of twist per unit length τ of the petal can be defined by the expression

$$r\tau_c = r\frac{\overline{\theta} - \theta}{h} = \frac{r}{\overline{r}}\left(\frac{1+\xi}{\tan\alpha} - \Delta\alpha\right) - \frac{1}{\tan\alpha} \quad (47)$$

That is,

$$r\tau_c = \left[\frac{\upsilon \sin\alpha \cos^2\alpha}{4(r/R)^2(1+\upsilon)} - \frac{\cos\alpha}{\sin\alpha}\right]\frac{\Delta r}{r} + \left[\frac{\cos^2\alpha}{4(r/R)^2}\left(\frac{1-2\sin^2\alpha}{1+\upsilon} + 2\sin^2\alpha\right) - 1\right]\Delta\alpha \quad (48)$$

At the two ends of the petal per twist pitch, the rotation is zero, and the compression strain obeys: $\varepsilon = -\varepsilon_0$; then, Eq. (46) and Eq. (48) become

$$\begin{cases} -\varepsilon_0 = \dfrac{\upsilon \sin^2\alpha \cos^2\alpha}{4(r/R)^2(1+\upsilon)}\dfrac{\Delta r}{r} + \left[\dfrac{\cos\alpha}{\sin\alpha} + \dfrac{\sin\alpha\cos\alpha}{4(r/R)^2}\left(\dfrac{1-2\sin^2\alpha}{1+\upsilon} + 2\sin^2\alpha\right)\right]\Delta\alpha \\ 0 = \left[\dfrac{\upsilon \sin\alpha \cos^2\alpha}{4(r/R)^2(1+\upsilon)} - \dfrac{\cos\alpha}{\sin\alpha}\right]\dfrac{\Delta r}{r} + \left[\dfrac{\cos^2\alpha}{4(r/R)^2}\left(\dfrac{1-2\sin^2\alpha}{1+\upsilon} + 2\sin^2\alpha\right) - 1\right]\Delta\alpha \end{cases} \quad (49)$$

The relation between axial strain ε_0 and transverse strain $\varepsilon_{Trans} = \frac{\Delta r}{r}$ can be expressed as

$$\varepsilon_{Trans} = \frac{C_4}{C_2C_3 - C_1C_4}\varepsilon_0 \quad (50)$$

where $C_1 = \frac{\upsilon \sin^2\alpha \cos^2\alpha}{4(r/R)^2(1+\upsilon)}$, $C_2 = \frac{\cos\alpha}{\sin\alpha} + \frac{\sin\alpha\cos\alpha}{4(r/R)^2}\left(\frac{1-2\sin^2\alpha}{1+\upsilon} + 2\sin^2\alpha\right)$,
$C_3 = \frac{\upsilon \sin\alpha \cos^2\alpha}{4(r/R)^2(1+\upsilon)} - \frac{\cos\alpha}{\sin\alpha}$, $C_4 = \frac{\cos^2\alpha}{4(r/R)^2}\left(\frac{1-2\sin^2\alpha}{1+\upsilon} + 2\sin^2\alpha\right) - 1$.

It is found that the coefficient between transverse and axial strains is affected by the helical angle α and Poisson's ratio υ. When the twist pitch of the fifth stage cable is of 427, 450, and 476 mm, the corresponding helical angle of the petal equals to 80, 80.5, and 81°, respectively. Substituting these values into Eq. (50), one can see that the axial strain of the cable ε_0 has a linear relationship with the transverse strain ε_{Trans}, which is displayed in **Figure 14**. We carried out an experiment by

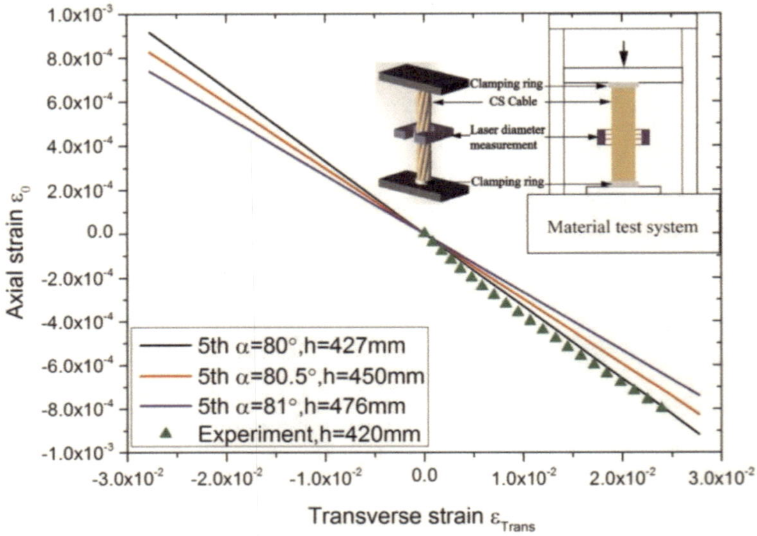

Figure 14.
Relationship between the axial strain ε_0 and the transverse strain ε_T. The colored lines are obtained by using the presented theoretical model. The green triangle symbols denote the experimental results.

compressing a CS cable which was fabricated in CASIPP to validate the analytical model. When the CS cable specimen is compressed along axial direction, it extends transversely with high resolution. The transverse extension can be measured by using a laser sensor. The experimental setup is schematically illustrated in the inset of **Figure 14**. One can find that the theoretical model shows perfect consistency with the experiment.

In **Figure 14**, we can get that the CS conductors with baseline and LTP (long twist pitch) shows a Tcs degradation after EM cycles, while for the samples with

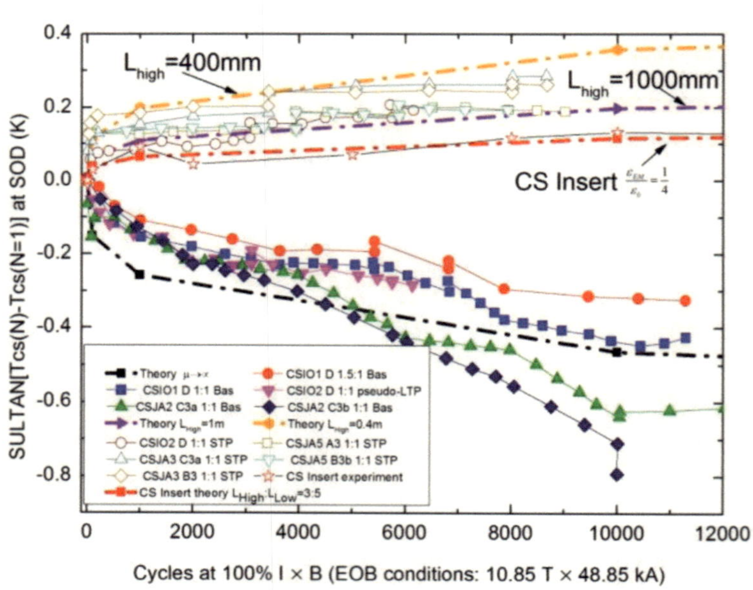

Figure 15.
A comparison of the experimental results and theoretical results based on the presented model [18].

STP (short twist pitch) shows an opposite phenomenon, the Tcs becomes a constant even have an enhanced. These experimental results presented in Ref. [30] have a contrast with analytical prediction for many years. Using the theoretical model proposed in this chapter, the Tcs enhancement and degradation behaviors can be predicted quantitatively. **Figure 15** shows a comparison of calculated results and experimental results, where the lines show the theoretical results and the dots are the experimental results. It is found that during the primary stage of the EM cycle there was a quick increase of axial strain that will lead to the Tcs drop dramatically. With the same reason, samples with LTP and baseline show an obviously degradation; this same rule applies to TF conductors. However, several STP CS conductors have an increase of Tcs, which means that the initial axial compression strain has been released by the EM loads, squeezing the cables at the high-field zone into the low-field region.

While it is easy to imagine when the expansions of the cable in the high-field zone get stacked, the strain will not be released and the Tcs will drop (see the black dash line in **Figure 15**). If the inner part of the conductor is smooth enough, the axial compression strain is released completely, the Tcs reaches its upper boundary which is shown by the orange dash line with circle symbol corresponding to the length of the high-field region is 400 mm. If the high-field region's length is 1000 mm, then the lower boundary of the enhancing Tcs is given with purple line with triangle symbol. We can find that the theoretical results agree with the experimental results very well.

6. Discussion

In the past half-century, the structure of CICC conductor was under continuous optimization and improvement. The CICC conductors have so many advantages such as good self-support, high operational stability, high current carrying capacity, low AC loss, etc., and they are widely used in the superconducting magnets. The nuclear fusion device CFTER built by China has also chosen the CICC conductors. However, there still have some problems to be solved: (1) there is a necessity for theoretical model to explain the degradation of critical current caused by mechanical deformation for the Nb3Sn strand; (2) accurate description of the mechanical response of complex strand structures in the coupling fields remains a challenging problem; (3) the untwisting behavior during the cable penetration is still not suppressed completely. One needs a more effective model to optimize the manufacturing process; and (4) the long-term stability and real-time monitoring of superconducting magnets are also a challenge for the engineers. Based on this, the equivalent mechanical parameters of CICC conductors and their mechanical behavior under coupled fields will be further studied. The authors hope that these models can provide a valuable reference for the related researchers.

Acknowledgements

This work is supported by the Funds of the National Natural Science Foundation of China (Nos. 11622217, 11802291, and 11872196), the National Key Project of Scientific Instrument and Equipment Development (11327802), and National Program for Special Support of Top-Notch Young Professionals. This work is also supported by the Fundamental Research Funds for the Central Universities (lzujbky-2017-ot18 and lzujbky-2017-k18).

Author details

Donghua Yue[1,2,3], Xingyi Zhang[1,2]* and Youhe Zhou[1,2]

1 Department of Mechanics and Engineering Sciences, College of Civil Engineering and Mechanics, Lanzhou University, Lanzhou, China

2 Key Laboratory of Mechanics on Disaster and Environment in Western China Attached to the Ministry of Education of China, Lanzhou University, Lanzhou, P.R. China

3 Institute of Systems Engineering, China Academy of Engineering Physics, Mianyang, China

*Address all correspondence to: zhangxingyi@lzu.edu.cn

IntechOpen

© 2018 The Author(s). Licensee IntechOpen. This chapter is distributed under the terms of the Creative Commons Attribution License (http://creativecommons.org/licenses/by/3.0), which permits unrestricted use, distribution, and reproduction in any medium, provided the original work is properly cited.

References

[1] Motojima O. The ITER project construction status. Nuclear Fusion. 2015;**55**(10):104023

[2] Mitchell N, Bessette D, Gallix R, et al. The ITER magnet system. IEEE Transactions on Applied Superconductivity. 2008;**18**(2):435-440

[3] File J, Martin GD, Mills RG, et al. Stabilized, levitated superconducting rings. Journal of Applied Physics. 1968; **39**(6):2623-2626

[4] Hoenig MO. Internally cooled cabled superconductors. Cryogenics. 1980; **20**(7):373-389

[5] Lubell MS, Chandrasekhar BS, Mallick GT. Degradation and flux jumping in solenoids of heat-treated Nb-25% Zr wire. Applied Physics Letters. 1963;**3**(5):79-80

[6] Hoenig M, Montgomery D. Dense supercritical-helium cooled superconductors for large high field stabilized magnets. IEEE Transactions on Magnetics. 1975;**11**(2):569-572

[7] Lue J. Review of stability experiments on cable-in-conduit conductors. Cryogenics. 1994;**34**(10): 779-786

[8] Ekin J. Strain scaling law for flux pinning in practical superconductors. Part 1: Basic relationship and application to Nb3Sn conductors. Cryogenics. 1980; **20**(11):611-624

[9] Nijhuis A. A solution for transverse load degradation in ITER Nb3Sn CICCs: Verification of cabling effect on Lorentz force response. Superconductor Science and Technology. 2008;**21**(5):054011

[10] Zhu JY, Luo W, Zhou YH, et al. Contact mechanical characteristics of Nb3Sn strands under transverse electromagnetic loads in the CICC cross-section. Superconductor Science and Technology. 2012;**25**(12):125011

[11] Feng J. A cable twisting model and its application in CSIC multi-stage cabling structure. Fusion Engineering and Design. 2009;**84**(12):2084-2092

[12] Qin J, Wu Y, Warnet LL, et al. A novel numerical mechanical model for the stress–strain distribution in superconducting cable-in-conduit conductors. Superconductor Science and Technology. 2011;**24**(6):065012

[13] Qin J-G, Xue T-J, Liu B, et al. Cabling Technology of Nb3Sn conductor for CFETR central solenoid model coil. IEEE Transactions on Applied Superconductivity. 2016; **26**(3):1-5

[14] Qin J, Dai C, Liu B, et al. Optimization of CFETR CSMC cabling based on numerical modeling and experiments. Superconductor Science & Technology. 2015;**28**(12):125008

[15] Yue D, Zhang X, Yong H, Zhou J, Zhou YH. Controllable rectification of the axial expansion in the thermally driven artificial muscle. Applied Physics Letters. 2015;**107**(11):111903

[16] Donghua Y, Xingyi Z, Youhe Z. Buckling behavior of Nb3Sn strand caused by electromagnetic force and thermal mismatch in ITER cable-in-conduit conductor. IEEE Transactions on Applied Superconductivity. 2017; **27**(7):1

[17] Qin J, Yee D, Zhang X, Wu Y, Liu X, Liu H, et al. Rotation analysis on large complex superconducting cables based on numerical modeling and experiments. Superconductor Science and Technology. 2018;**31**(2):025001

[18] Yue D, Zhang X, Zhou Y-H. Theoretical analysis for the mechanical behavior caused by an electromagnetic cycle in ITER Nb3Sn cable-in-conduit conductors. Acta Mechanica Sinica. 2018;**34**(4):614-622

[19] Costello GA. Theory of Wire Rope. New York: Springer Science & Business Media; 1997

[20] Nabara Y, Suwa T, Takahashi Y, et al. Behavior of cable assembled with conduit for ITER central solenoid. IEEE Transactions on Applied Superconductivity. 2015;**25**(3):1-5

[21] Hetenyi M. Beams on Elastic Foundation: Theory with Applications in the Fields of Civil and Mechanical Engineering. 1971

[22] Niu K. Compressive Behavior of Sandwich Panels and Laminates with Damage, Georgia Institute of Technology, Atlanta. 1998

[23] Niu K, Talreja R. Buckling of a thin face layer on Winkler foundation with debonds. Journal of Sandwich Structures and Materials. 1999;**1**(4): 259-278

[24] Mitchell N, Devred A, Larbalestier DC, et al. Reversible and irreversible mechanical effects in real cable-in-conduit conductors. Superconductor Science and Technology. 2013;**26**(11): 114004

[25] Li Y, Ta W, Gao Y. Prediction for IC degradation of superconducting strand under bending. Cryogenics. 2013;**58**: 20-25

[26] Li Y, Yang T, Zhou Y, et al. Spring model for mechanical–electrical properties of CICC in cryogenic–electromagnetic environments. Cryogenics. 2014;**62**:14-30

[27] Hobbs RE. Pipeline buckling caused by axial loads. Journal of Constructional Steel Research. 1981;**1**(2):2-10

[28] March SA, Bruzzone P, Stepanov B, et al. Results of the TFEU6 sample tested in SULTAN. IEEE Transactions on Applied Superconductivity. 2013; **23**(3):4200204

[29] Martovetsky NN, Reiersen WT, Murdoch GR, Bruzzone P, Stepanov B. Qualification of the US conductors for ITER TF magnet system. ITPS. 2018; **46**(5):1477-1483

[30] Devred A, Backbier I, Bessette D, et al. Challenges and status of ITER conductor production. Superconductor Science and Technology. 2014;**27**(4): 044001